Adiabatic Fixed-Bed Reactors

Adiabatic Fixed-Bed Reactors
Practical Guides in Chemical Engineering

Jonathan Worstell
Professor of Chemical Engineering
University of Houston

President
Worstell and Worstell, Consultants

AMSTERDAM • BOSTON • HEIDELBERG • LONDON
NEW YORK • OXFORD • PARIS • SAN DIEGO
SAN FRANCISCO • SINGAPORE • SYDNEY • TOKYO
Butterworth-Heinemann is an imprint of Elsevier

Butterworth-Heinemann is an imprint of Elsevier
225 Wyman Street, Waltham, MA 02451, USA
The Boulevard, Langford Lane, Kidlington, Oxford OX5 1GB, UK

Notices
Knowledge and best practice in this field are constantly changing. As new research and
experience broaden our understanding, changes in research methods, professional practices,
or medical treatment may become necessary.

Practitioners and researchers must always rely on their own experience and knowledge in
evaluating and using any information, methods, compounds, or experiments described herein.
In using such information or methods they should be mindful of their own safety and the safety
of others, including parties for whom they have a professional responsibility.

To the fullest extent of the law, neither the Publisher nor the authors, contributors, or editors,
assume any liability for any injury and/or damage to persons or property as a matter of products
liability, negligence or otherwise, or from any use or operation of any methods, products,
instructions, or ideas contained in the material herein.

Library of Congress Cataloging-in-Publication Data
A catalogue record for this book is available from the Library of Congress.

British Library Cataloguing-in-Publication Data
A catalogue record for this book is available from the British Library.

ISBN: 978-0-12-801306-9

For information on all Butterworth-Heinemann
publications visit our website at http://store.elsevier.com

Working together
to grow libraries in
developing countries

ELSEVIER   Book Aid
International

www.elsevier.com • www.bookaid.org

Butterworth-Heinemann is an imprint of Elsevier
225 Wyman Street, Waltham, MA 02451, USA
The Boulevard, Langford Lane, Kidlington, Oxford OX5 1GB, UK

Notices
Knowledge and best practice in this field are constantly changing. As new research and
experience broaden our understanding, changes in research methods, professional practices, or
medical treatment may become necessary.

Practitioners and researchers must always rely on their own experience and knowledge in
evaluating and using any information, methods, compounds, or experiments described herein.
In using such information or methods they should be mindful of their own safety and the safety
of others, including parties for whom they have a professional responsibility.

To the fullest extent of the law, neither the Publisher nor the authors, contributors, or editors,
assume any liability for any injury and/or damage to persons or property as a matter of products
liability, negligence or otherwise, or from any use or operation of any methods, products,
instructions, or ideas contained in the material herein.

Library of Congress Cataloging-in-Publication Data
A catalog record for this book is available from the Library of Congress.

British Library Cataloguing-in-Publication Data
A catalogue record for this book is available from the British Library.

ISBN: 978-0-12-801306-9

For information on all Butterworth-Heinemann publications
visit our website at http://store.elsevier.com

This book has been manufactured using Print On Demand technology. Each copy is produced to
order and is limited to black ink. The online version of this book will show color figures where
appropriate.

Working together
to grow libraries in
developing countries

www.elsevier.com • www.bookaid.org

DEDICATION

To our sons

Jeffrey H. Worstell

John H. Worstell

CONTENTS

CHAPTER *1*

Introduction

1.1 SLOW REACTIONS AND CATALYSTS

Many reactions are too slow to be commercially viable, no matter how much we desire the product. Over the centuries, we have learned that adding a particular chemical, such reactions proceed at commercially viable rates. We call that particular chemical a "catalyst." A catalyst is a substance that accelerates a specific reaction toward its equilibrium but remains unchanged after the reaction achieves equilibrium.[1] Many times the catalyst and product can be in the same phase at the end of the reaction. In other words, the catalyst contaminates the product and must be removed before the product is sold. Achieving this separation can be expensive, which increases the asking price for the product. Separation costs can actually make the product commercially unviable.

Two options exist for alleviating the problem of catalyst separation: (1) increase catalyst productivity to the point that only parts per million of it remain in the product and (2) make the catalyst stationary relative to product formation.

The first option may provide the simplest process, provided we can find a catalyst with sufficient productivity. Catalyst productivity is the rate at which reactant disappears or product appears

- per number of catalytic sites;
- per catalyst surface area;
- per catalyst volume or per catalyst weight.

The chemical processing industry (CPI) generally defines catalyst productivity as rate per unit catalyst weight.[2]

The chemical process could be as simple as charge the reactor with reactant, inject catalyst, stir, empty the reactor, then sell the product. While such a process certainly has its advantages, it also has some major disadvantages. First, since the catalyst remains in the product, its components must appear in the product specification, even if present as parts per million. If the product is a chemical intermediate, which means it undergoes further reaction, then any change in its specification must be communicated to its users and must gain their approval. In many cases, the processes using such chemical intermediates are sensitive to those parts per million of catalyst. In some cases, those parts per million are beneficial to the process; in most cases, they are a detriment to the process. Thus, the parts per million catalyst in the product must be controlled, which represents a manufacturing cost. A greater disadvantage is we cannot change the catalyst without informing our customers since we will have to alter the product specification. If we adopt a more cost-effective catalyst, our customers are going to want some or all of that cost savings, i.e., they will want a reduction in product price. Or—if those catalyst parts per million are beneficial to the customer, then the customer may simply refuse our request to change catalyst, thereby forcing us to supply current product for the life of the contract. But, the greater disadvantage of this option is the product tells the world what catalyst we used to manufacture product. Such insight allows them to use a similar catalyst or to develop a more productive catalyst.

Second, if the product enters the food chain, either as an ingestible or as a food container or wrapper, then it must undergo significant testing and costly certification by a variety of government agencies. Once a product gains regulatory approval, any change to its process or specification can represent recertification, which is a costly proposition. Thus, improving catalyst performance or changing the catalyst becomes problematic due to all the regulatory issues raised by such a change.

Third, what to do with product not meeting specification? This situation occurs when the catalyst is not as productive as it should be. In this case, the product contains too much catalyst. Diluting the "off-spec" material with "on-spec" product is the most likely action. Diluting in this fashion can require considerable time if the volume of off-spec material is large.

The second option is to immobilize the catalyst relative to process flow. In this case, the catalyst is stationary and the process fluid flows over it or through it. To immobilize a catalyst, we must adsorb, coprecipitate, or attach by some other means the active chemical component of the catalyst onto a solid. The solid must provide enough surface area to meet the productivity requirements of the process. In other words, the solid-supported catalyst must possess a productivity sufficient for economic viability. The solid must also possess enough strength to support its own weight when contained in a vessel. If these requirements are met, then we can install a fixed-bed reactor in our process.

There are several advantages to using a fixed-bed reactor in a catalyzed process. First, we know where the catalyst is—most of the time. Second, we do not have to recover the catalyst from any process stream or from our product. Third, product registration with a government agency may not be required, or, if required, will be simplified considerably since we do not leave trace amounts of catalyst in the product. This advantage can reduce the time to develop a given process and reduce development costs considerably since registration with a government agency may not be required. Fourth, we do not have to disclose the catalyst's components in our product specification since the product does not contain them. This advantage allows us to decide whether to patent our catalyst or use it as a trade secret. Also, we do not have to inform customers of any catalyst change, unless so stipulated by the sales contract. Fifth, this option expands the choice of catalyst since the catalyst does not have to possess extraordinarily high catalytic productivity. Sixth, solid catalysts can be regenerated. Regeneration is a procedure that returns catalytic activity to or close to its original value. Regeneration generally involves burning organic chemical deposit off the catalyst's solid surface, then chemically reducing the metal component of the catalyst before exposing it to process fluid.

Fixed-bed reactors also possess disadvantages. First, solid-supported catalytic productivity declines with time for a variety of reasons. In many cases, it is difficult to distinguish which mechanism is causing the catalytic deactivation. Thus, solid-supported catalysts require periodic regeneration or replacement. Second, solid-supported catalysts can move *en masse* when the screens retaining the catalyst bed fail. When the retaining screens fail, catalyst flows into downstream piping and equipment. Such a screen failure initiates a massive maintenance effort to rehabilitate the process. Third, the solid-supported catalyst can be crushed by its own weight, thereby plugging the fixed-bed reactor. Fourth, the catalyst can attrite, i.e., form dust or fines, which escape the reactor and enter the downstream process. Such fines can blind filters, plug control valves, or accumulate in process dead-legs. In general, such fines cause havoc in the process.

When evaluating the advantages and disadvantages of fixed-bed reactors, the former generally overpowers the latter, making fixed-bed reactors the most popular method for catalyzing reactions in gas or liquid process streams.[2(pp5−7)]

1.2 FIXED-BED REACTOR CLASSIFICATION

We generally classify fixed-bed reactors as adiabatic or nonadiabatic. Adiabatic and nonadiabatic are thermodynamic terms describing energy exchange. Energy is an extensive variable, thus it depends upon the quantity of material being investigated. We call a given quantity of material a "system" and we idealize it as separated by real or imaginary walls from the rest of the universe, which we call the "surroundings." By separating the system in this manner, we can control or observe energy exchange between it and its surroundings. A system may be isolated, in which case energy, either as heat or work, is not exchanged between the system and its surroundings. An adiabatic system exchanges work, but not heat, with its surroundings.[3] Thus, the walls of an adiabatic system are heat opaque; they are insulated. But, systems with no or poor insulation can be adiabatic if any change within them occurs more rapidly than does achieving a new energy equilibrium with their surroundings, which is the case for fixed-bed reactors: localized events occur, then disappear, within fixed-bed reactors before a new energy equilibrium can be established with their surroundings.[4] Adiabatic fixed-bed reactors are the reactor of choice for solid-supported catalysts due to their simple design, straightforward construction, and "hassle-free" operation.

Nonadiabatic fixed-bed reactors exist. They are used primarily in processes that are highly exothermic, such as oxidation of hydrocarbons. These reactors require heat transfer along the catalyst bed, which necessitates a high heat transfer surface area to reaction volume (HTSA/RV) ratio. Filling a small diameter pipe, i.e., a tube, with solid catalyst, then submerging it in a flowing fluid is the easiest method for achieving a high HTSA/RV ratio. However, one tube is commercially unviable, but multiple tubes in one container or shell can be commercially viable. Such fixed-bed reactors resemble upended heat exchangers. Multitube fixed-bed reactors are also used for endothermic reactions, steam reforming being an example.

1.3 PROCESSES OPERATING FIXED-BED REACTORS

1.3.1 Ammonia Synthesis

Civilization started with the advent of agriculture. As agriculture became more productive, the human population increased, thereby increasing the demand for agricultural products. The use of fertilizers was and is an integral component for increasing agricultural productivity. Nitrogen- and phosphorus-containing chemicals are the most important ingredients of a fertilizer. Plants use the nitrogen to synthesize purines and pyrimidines, which form the base pairs in plant DNA, and the phosphorus, as phosphate, forms the DNA backbone from which the purines and pyrimidines extend.

Until the third quarter of the nineteenth century, animal waste and decaying vegetation supplied agriculture with nitrogen-containing fertilizer. Guano from Chile and Peru provided most of the nitrogen-containing animal waste. Decaying vegetation came from plants rotting in fields. By mid-nineteenth century, demand for nitrogen-containing chemicals was beginning to overtake their supply. The burgeoning steel industry came to agriculture's rescue. Steel requires coke which comes from the destructive distillation of coal. Coal comes from plants; thus, nitrogen-containing chemicals are in it. The steel industry recovered those nitrogen-containing chemicals and sold them to agriculture as fertilizer. However, by 1870 or so, guano, decaying vegetation, and coke-derived nitrogen could not meet fertilizer demand. About 1870, chemical companies started mining saltpeter, then modified it for agricultural use. The fertilizer market stabilized until 1900. Around the turn of the century, reports began to appear claiming the global supply of saltpeter reserve was about 30 years. Shortly thereafter, chemists in Germany realized that air represented an unlimited supply of nitrogen for fertilizer. By 1914, Germany had developed an industry that converted atmospheric nitrogen into ammonia, which could be used as fertilizer.[5]

The synthesis of ammonia from nitrogen is highly exothermic:

$$N_{2(g)} + 3H_{2(g)} \Leftrightarrow 2NH_{3(g)}$$
$$\Delta H_{698°C} = -13.3 \, kcal/mol$$

While the above reaction is possible, its rate of approach to equilibrium between products and reactants is slow. To be commercially viable, the above reaction requires a catalyst. Fritz Haber, who performed the laboratory investigation for ammonia synthesis, used osmium as a catalyst. Osmium is relatively rare, thus expensive, and its oxides are highly volatile and toxic. Hence, Badische Anilin- und Soda-Fabrik, now BASF, which owned the rights to Haber's laboratory work, initiated a project in 1910 to find a cheaper, more suitable catalyst for their ammonia process. Alwin Mittasch and other chemists at BASF found that iron would catalyze the above reaction. Carl Bosch developed the BASF ammonia process using iron as its catalyst.[5] Iron has been the catalyst of choice for ammonia synthesis since that discovery.

Magnetite forms the base of today's ammonia process catalysts. Magnetite, containing trace quantities of potassium oxide, calcium oxide, magnesium oxide, alumina, and silica, is preferred. The iron catalyst forms upon reducing magnetite. The trace oxides are also reduced and volatilized, thereby forming pores in the reduced iron. These pores increase the iron's catalytic surface area.

Since ammonia formation is highly exothermic, the fixed-bed reactor for the process must have a high HTSA/RV ratio, which is accomplished by inserting a steel cylinder into a high-pressure shell. Within the steel cylinder are several catalyst beds, each resting on a screen and separated by an open space from the next catalyst bed. This design forms an annular flow path between the inserted steel cylinder and the high-pressure reactor shell. Feed gases enter the top or bottom of the high-pressure shell, depending upon reactor design. For this discussion, we assume the feed gases enter the top of the reactor shell. The feed gases then flow down the annulus to the bottom of the reactor shell, where they change flow direction, pass around a bundle of tubes containing hot product that flows countercurrent to the feed gas, then the feed gases rise through a centerline pipe to the top of the inserted cylinder. The feed gases again reverse direction at the top of the inserted cylinder and begin flowing downward through the first catalyst bed. The first catalyst bed is sized to achieve a given outlet temperature. In other words, the reactant gases and product gases exit the first catalyst bed at a predetermined temperature. Cold feed gases are injected into the space between the first catalyst bed and the second catalyst bed. This cold feed gas reduces the temperature of the gas mixture before it enters the second catalyst bed and it dilutes the product concentration, thereby utilizing Le Chatelier's principle. This gas mixture then enters the second catalyst bed, which is sized to yield a predetermined gas discharge temperature. Cold feed gas injected between the second and third catalyst beds quenches the temperature of the gas mixture before it enters the third catalyst bed. This sequence of events continues through the length of the steel cylinder. After exiting the bottom catalyst bed, the hot gas enters the tube bundle at the bottom of the reactor shell. Outside these tubes, feed gas circulates, heats, and changes flow direction, then enters the centerline pipe and rises to the top of the steel cylinder. In other words, the discharging product gases heat the incoming feed gases. The product gases exit the tube bundle and enter the discharge nozzle of the reactor shell. In summary, this design contains an internal heat exchanger and a series of catalyst beds sized to produce a given temperature profile within one reactor shell. This design is common for highly exothermic, gas phase reactions.

1.3.2 Oxidation of Ethylene

The oxidation of ethylene to form ethylene oxide is another highly exothermic reaction requiring a catalyst. At least four variations of the process are used commercially; however, they all employ a silver impregnated solid-supported catalyst as well as air or oxygen to oxidize ethylene. The solid support is alumina spheres or extruded alumina. The latter is wet alumina—which resembles wheat dough—that has been forced through a die. The die shapes the extruded alumina into a variety of configurations, e.g., cylinders, trilobes, hollow cylinders, or rings. The extruded alumina is then calcined to set its pore structure and surface area. Essentially, it is a pasta-making process.

The formal reaction for ethylene oxide is

$$2CH_2 = CH_2 + O_2 \rightarrow 2(CH_2CH_2)O$$
$$\Delta H = -105 \text{ kJ/mol}$$

Thus, the fixed-bed reactor requires a high HTSA/RV ratio, which is achieved by using a reactor containing multiple tubes or flow paths. Each tube is 6−15 m long and has a 20 mm to 50 mm inner diameter. A multitube reactor will contain many hundred, if not several thousand, tubes. Such reactors resemble upended heat exchangers. On the "tube side," i.e., inside each tube, ethylene and air or oxygen at 200−300°C and 1−3 MPa flow downward, over and through the porous, solid-supported catalyst. On the "shell side," i.e., exterior to the tubes, coolant flows upward. The coolant may be a refinery product, such as diesel fuel or some other oil or it may be a heat transfer fluid such as Dowtherm or Syltherm. As the solid-supported catalyst ages, its selectivity declines, thereby producing more by-products, which produce more heat than does the desired reaction.

Charging catalyst to a multitube reactor requires significant technical skill; otherwise, "channeling" occurs. Channeling happens when one or more tubes have a lower pressure differential top to bottom than the remaining tubes. Since fluids follow the path of least resistance, more fluid flows through those tubes possessing a lower pressure differential than through those tubes possessing a higher pressure differential. Catalyst activity declines faster in the former tubes than in the latter tubes. This decline adversely impacts the performance of the fixed-bed reactor since more reactant flows through the lower pressure differential tubes than through the higher pressure differential tubes. Thus, the importance of filling each tube uniformly, then measuring its pressure differential.

1.3.3 Catalytic Reforming
Some processes require maximum contact time between the reactants and the solid-supported catalyst while minimizing the inlet to outlet pressure differential. One such process is catalytic reforming.

The quest for fuel efficiency has led to development of high compression internal combustion engines. High compression engines require fuels that do not predetonate, i.e., that do not "knock." Knocking occurs in an internal combustion engine when the fuel auto-ignites ahead of the spark-initiated flame front. Such ignition establishes a shock wave within the engine cylinder, thereby causing a sharp pressure rise that exceeds the engine design specification. This pressure increase damages the engine.

During the 1920s, various researchers demonstrated that branched hydrocarbons caused considerably less engine knock and that highly branched hydrocarbons tended not to knock at all. Petroleum refiners, therefore, developed the "reforming" process, a process designed to reform straight-chain hydrocarbons into branched hydrocarbons. They essentially modified their existing thermal cracking processes. Refiners increased process temperature from 400°C to 650°C and they increased process pressure from 100 to 1000 psig. The new process conditions produced a gasoline with moderate antiknock characteristics. The first dedicated reforming process streamed in 1930.

Thermal reforming is a free radical process. The free radical process creates significant amounts of olefin gases and polymeric "tars." The olefin gases led to development of the petrochemical industry. The polymeric tars caused problems in the reforming process.

By the late 1930s, the compression ratios of automobile engines, in general, and airplane engines, in particular, required a gasoline possessing better antiknock characteristics than what the thermal reforming process could deliver. And, with the start of the Second World War, demand for antiknock gasoline far exceeded what the thermal reforming process could produce. Thus refiners switched to catalytic reforming.

Two basic reforming catalysts exist: solid acids and solid acids impregnated with as much as 1% platinum. Catalytic reforming occurs via the formation of carbonium ions rather than free radicals. Because catalytic reforming occurs via a carbonium ion formation, it produces much less olefin gas, if any, and the solid acid collects the polymer tars as deposit, i.e., as "coke." Combustion regeneration of the catalyst removes the coke.

Catalytic reforming is a high throughput process. Thus, catalytic reforming reactors require minimal inlet to discharge pressure differential. The radial flow fixed-bed reactor meets these process needs. A basket through which vapor passes contains the solid-supported catalyst. The catalyst-filled basket forms a cylindrical annulus within the reactor shell. Reactant vapor enters the annulus, then flows radially inward through the catalyst mass to the reactor's centerline along which it descends to the discharge port. Or, the reactant enters the reactor through a centerline nozzle, then flows radially outward through the catalyst mass to the annular flow path along which it descends to the discharge port. The product gases are then separated and the hydrogen recycles to the fixed-bed reactor.

1.3.4 Steam Methane Reforming
Several chemical processes require large quantities of high-purity hydrogen. Ammonia synthesis and petroleum hydrocracking being two such processes. Hydrocracking combines catalytic cracking and hydrogenation to produce more gasoline from each barrel of refined crude oil. Refiners also use considerable hydrogen to treat various refinery streams, thereby upgrading the product of the stream and improving the feed quality of the stream for subsequent processing.

Refiners recover significant amounts of hydrogen from their catalytic reformer off-gas. However, when the hydrogen from catalytic reforming is insufficient, a refinery must develop an independent hydrogen source. Steam methane reforming provides that independent source of hydrogen in the United States, where natural gas is plentiful. Steam naphtha reforming provides independently sourced hydrogen outside the United States where liquid hydrocarbon is more plentiful than natural gas.

The steam methane reforming reaction is

$$CH_4 + H_2O \Leftrightarrow CO + 3H_2$$
$$CO + H_2O \Leftrightarrow CO_2 + H_2$$

The first reaction is highly endothermic; thus, it requires heat. The second reaction is mildly exothermic. Steam methane reforming requires a high HTSA/RV ratio, not for removing heat from the reactor, but for adding heat to the reactor. As discussed above, multiple tube reactors provide high HTSA/RV ratios.

Steam methane reforming occurs in reactors that are actually tubular furnaces. A typical steam methane reformer preheats the gas feed in a convection furnace box, then injects the hot gas into vertical tubes placed in a radiation furnace box. The vertical tubes contain a nickel impregnated solid-supported catalyst. The catalyst-filled tubes are designed for a specified heat flux along the tube.

1.3.5 Double-Bond Isomerization/Metathesis/Hydrogenation

Many catalyzed reactions are slightly endothermic or exothermic; therefore, when feed encounters the catalyst, product begins to form at process temperatures. Double-bond isomerization using a solid acid catalyst or using a solid base catalyst is an example of a slightly exothermic reaction. Mild hydrogenation is another example of a catalyzed reaction that occurs at process temperatures.

Some reactions are thermally neutral. Double-bond metathesis is one such reaction. In double-bond metathesis, carbon–carbon double bonds are broken, thereby releasing energy, and other, different carbon–carbon double bonds are formed from the resulting fragments, thereby utilizing the released energy.

1.4 OPERATION OF FIXED-BED REACTORS

For mildly endothermic or exothermic reactions, petrochemical companies typically use large, vertical, cylindrical reactors filled with solid-supported catalyst. These reactors operate adiabatically. As always, there is an exception for a process requiring high flow rate and minimal pressure differential across the catalyst mass. For such processes, large diameter, vertically short fixed-bed reactors are preferred. In such cases, spherical reactors with horizontal catalyst layers across their diameter are not uncommon.

For gas processes, feed enters the top of the reactor and flows downward through the catalyst mass. For liquid processes where fluidization is not an issue, feed enters the bottom of the reactor and flows upward through the catalyst mass. In either case, a screen protects the bottom nozzle of the reactor. This screen is generally covered with unreactive ceramic spheres. Large spheres contact the screen; layers of progressively smaller spheres cover the large spheres. For gas processes, these spheres provide a filter to trap small entrained catalyst particles before they reach the bottom screen. For upflow liquid processes, these spheres disperse the inlet flow across the diameter of the reactor, thereby minimizing channeling through the catalyst mass.

A similar mass of ceramic spheres lies atop the catalyst mass. Small diameter spheres rest directly upon the catalyst mass. Layers of progressively larger spheres cover the layer of small spheres. These spheres disperse the feed in gas processes and act as a filter for catalyst fines in liquid processes.

1.5 SUMMARY

This chapter discussed the concept of catalysis and the reasons for designing, building, and operating fixed-bed reactors. It also presented the types of fixed-bed reactors available for use and related reactor type to the heat transfer requirements of the chemical process. This chapter discussed examples of fixed-bed reactors currently in use in the CPI. It also introduced the classification of fixed-bed reactors as adiabatic or nonadiabatic. This chapter also introduced the concepts of catalytic productivity and selectivity.

REFERENCES

1. Gates B. *Catalytic chemistry*. New York, NY: John Wiley & Sons; 1992. p. 2

2. Satterfield C. *Heterogeneous catalysis in practice*. New York, NY: McGraw-Hill Book Company; 1980. p. 44

3. Steiner L. *Introduction to chemical thermodynamics*. New York, NY: McGraw-Hill Book Company; 1941. p. 26

4. Paul M. *Principles of chemical thermodynamics*. New York, NY: McGraw-Hill Book Company; 1951. p. 4–5

5. Stoltzenberg D. *Fritz Haber: chemist, Nobel laureate, German, Jew—a biography*. Philadelphia, PA: Chemical Heritage Press; 2004 [chapter 5]

CHAPTER 2

Fundamentals of Fixed-Bed Reactors

2.1 ANALYSIS OF FIXED-BED REACTORS

Fixed-bed reactors come in all sizes, but their shape is generally restricted to that of a circular cylinder which is later filled with solid-supported catalyst. Feed enters one end of the reactor and product exits the other end of the reactor. The component mass balance for a fixed-bed reactor is

$$\frac{\partial C_A}{\partial t} + \left(v_r \frac{\partial C_A}{\partial r} + \frac{v_\theta}{r} \frac{\partial C_A}{\partial \theta} + v_z \frac{\partial C_A}{\partial z} \right)$$
$$= D_{AB} \left(\frac{1}{r} \frac{\partial}{\partial r} \left(r \frac{\partial C_A}{\partial r} \right) + \frac{1}{r^2} \frac{\partial^2 C_A}{\partial \theta^2} + \frac{\partial^2 C_A}{\partial z^2} \right) + R_A$$

where C_A is the concentration of component A in moles/m^3. We operate fixed-bed reactors at constant feed and product flow rates; thus, they operate at steady state, which means $(\partial C_A/\partial t) = 0$. At high feed rates, the flow through a fixed-bed reactor, especially one filled with small solid-supported catalyst pellets or extrudates, behaves in a plug flow manner. Therefore, $v_r = v_\theta = 0$. The reactor design, catalyst size, and fluid flow rate combine to determine whether the dispersion terms impact the performance of a fixed-bed reactor; the dispersion terms are

$$\frac{1}{r} \frac{\partial}{\partial r} \left(r \frac{\partial C_A}{\partial r} \right) + \frac{1}{r^2} \frac{\partial^2 C_A}{\partial \theta^2} + \frac{\partial^2 C_A}{\partial z^2}$$

The Peclet number for the process and the various aspect ratios of the fixed-bed reactor determine the impact of the dispersion terms. The mass Peclet number quantifies the ratio of bulk mass transport to diffusive mass transport. We define the mass Peclet number as

$$Pe = \frac{Lv}{D_{AB}}$$

where L is a characteristic length; v is fluid velocity; and D_{AB} the diffusivity constant for component A in bulk component B. L can be the height of the catalyst mass, identified as Z, or the diameter of the reactor, identified as D. The product of the directional aspect ratio and the Peclet number determines the importance of dispersion in that direction. If $(Z/D_p) * Pe$ is large, where D_p is the diameter of the solid-supported catalyst, then dispersion in the axial direction; i.e., along the z-axis, is negligible. Generally, Z is large for a fixed-bed reactor and D_p is small; thus, we can confidently neglect the $\partial^2 C_A/\partial z^2$ dispersion term. With regard to the axial aspect ratio itself, Carberry[1] purports that $(Z/D_p) \geq 150$ ensures no axial dispersion in a fixed-bed reactor. In the radial direction, we seek the opposite outcome because conduction is the only mass transfer mechanism in that direction. Thus, we want a small radial Pe number, which is Dv/D_{AB}. It is heat transfer at the fixed-bed reactor wall that induces radial conduction and mass transfer. Hence, radial dispersion can only be neglected for adiabatic reactors. However, Carberry[2] writes that radial aspect ratios R/D_p of 3 to 4 ensures negligible radial dispersion, R being the reactor's radius. Thus, $\partial C_A/\partial r$ is zero. Chemical engineers generally neglect dispersion in the azimuthal direction. Therefore, the above component mass balance reduces to

$$v_z \frac{dC_A}{dz} = R_A$$

where R_A represents reactant consumption.

We generally do not characterize fixed-bed reactor operation by its fluid velocity. Instead, we characterize fixed-bed operation by its volumetric flow rate. Rearranging the above equation and multiplying by the cross-sectional area of the reactor, we obtain

$$v_z A \, dC_A = R_A A \, dz$$

Note that $v_z A$ is volumetric flow rate and $A \, dz$ is dV, the differential volume of catalyst. V is the fluid volume of the catalyst mass, which is the true volume of the reactor. However, the CPI generally identifies V as the weight W of the solid-supported catalyst charged into the reactor shell. Using W instead of V complicates our analysis for a fixed-bed reactor. Therefore, we will identify V as the true volume for a fixed-bed reactor. That volume is

$$\frac{W_{Catalyst}}{\rho_{LBD}} = V_{Catalyst}$$

where $W_{Catalyst}$ is the weight of catalyst charged into the reactor shell and ρ_{LBD} is the loose bulk density of the catalyst. We use loose bulk density because we do not shake a commercial reactor while filling it. We may shake a laboratory reactor tube or shell while filling it, but we generally do not shake commercial reactors when filling them. Note that $V_{Catalyst}$ includes fluid volume as well as solid volume. We are interested in the fluid volume of the catalyst mass, which is

$$V_{Fluid} = \varepsilon V_{Catalyst}$$

where ε is the void fraction of the solid-supported catalyst.

Thus, the above equation becomes

$$Q \, dC_A = R_A \, dV_{Fluid}$$

Rearranging this last equation gives us

$$\frac{dC_A}{R_A} = \frac{dV_{Fluid}}{Q}$$

The boundary conditions for this equation are $C_A = C_{A,In}$ at $V = 0$, and $C_A = C_{A,Out}$ at $V = V$. Integrating with these boundary conditions yields

$$\int_{C_{A,In}}^{C_{A,Out}} \frac{dC_A}{R_A} = \int_0^V \frac{dV_{Fluid}}{Q} = \frac{V_{Fluid}}{Q}$$

V_{Fluid}/Q has units of time, generally in minutes or seconds. We call V_{Fluid}/Q space time and it represents the average time required to traverse a given flow path from the leading edge of the catalyst mass to the trailing edge of the catalyst mass. We cannot as yet integrate the left-hand side of the above equation because we have not specified R_A mathematically. To do so, we must know what happens within the catalyst mass during reactor operation.

But, just what does occur within the catalyst mass during reactor operation? First, fluid enters one of the myriad flow tubes or channels passing through the catalyst mass. These flow channels form between catalyst pellets or extrudates. The bulk fluid moves through the catalyst mass via these flow tubes or channels.

Second, whenever a fluid flows over a solid or liquid surface, a stagnant film forms along that surface. At the surface, fluid velocity in the direction of bulk flow is zero. At the outer edge of the stagnant film, fluid velocity is that of the bulk fluid. Within the stagnant film, a velocity gradient in the direction of bulk fluid flow exists. However, there is no fluid velocity component normal, i.e., perpendicular to the solid or liquid surface. In other words, convection does not occur across the stagnant film. Thus, reactant and product molecules diffuse across the stagnant film.

Third, most solid-supported catalysts are porous, which greatly increases their surface area. Increasing surface area leads to an increased number of catalytic sites available for reaction, thereby increasing catalyst productivity. Reactant molecules migrate along these pores via diffusion. When they encounter an empty catalyst site, they become product. These product molecules must then diffuse through the pore network of the solid-supported catalyst, diffuse across the stagnant film surrounding each catalyst pellet or extrudate, then enter the bulk fluid to exit the catalyst mass. The concentration difference between a feed sample and a product sample represents the sum of these mechanisms.

Figure 2.1 presents a schematic of catalysis in a porous solid. The reactant concentration at the outer boundary of the stagnant film is the concentration of reactant in the feed, i.e., it is C_{BF} in moles/m^3.

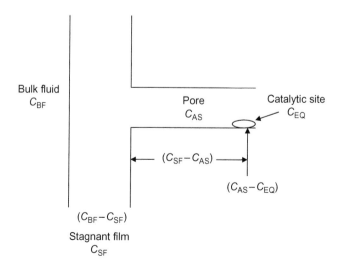

Figure 2.1 Schematic of adsorption impedances.

Reactant molecules move across the stagnant film by molecular diffusion, which we generally model as a linear concentration difference. That difference is $C_{BF} - C_{SF}$, where C_{SF} is the concentration of reactant at the surface of the catalyst in moles/m^3. Reactant then diffuses from the surface of the catalyst along pores to the catalytic sites inside the solid. Reactant movement within the pore is also by molecular diffusion, which we model as a linear concentration difference. Catalytic sites occur along the length of the pore, thus reactant concentration changes along the length of the pore. Reactant concentration at a given catalyst site is C_{AS} in moles/m^3. Thus, the concentration difference to that point in the pore is $C_{SF} - C_{AS}$. Equilibrium may be established at the catalytic site; equilibrium concentration is C_{Eq} in moles/m^3.

We can write the rate of each of the above described mechanisms as a concentration difference. The reactant conversion rate is

$$R_{Rxn} = k_{Rxn}(C_{AS} - C_{Eq}) \qquad (2.1)$$

where k_{Rxn} is the reaction rate constant at the catalytic site. k_{Rxn} has units of 1/minutes (min) or 1/seconds (s). The rate of reactant movement along the pore is

$$R_{PD} = k_{PD}(A_P/V_P)(C_{SF} - C_{AS}) \qquad (2.2)$$

where k_{PD} is the pore diffusion mass transfer rate constant (m/s), A_P is the average cross-sectional area of a pore (m^2), and V_P is average pore volume (m^3). The rate of reactant movement through the stagnant film surrounding the catalyst pellet is

$$R_{SFD} = k_{SFD}(S_{Film}/V_{Film})(C_{BF} - C_{SF}) \qquad (2.3)$$

where k_{SFD} is the stagnant film mass transfer rate constant (m/s), S_{Film} is the surface area (m^2), and V_{Film} is the volume (m^3) of the stagnant film surrounding the catalyst pellet or extrudate.

The only reactant concentrations known with any accuracy are C_{BF} and C_{Eq}. Thus, the mathematical expression for the overall rate of reactant conversion must be in terms of C_{BF} and C_{Eq}. Solving Eq. (2.1) for C_{AS} yields

$$\frac{R_{Rxn}}{k_{Rxn}} + C_{Eq} = C_{AS} \qquad (2.4)$$

Solving Eq. (2.2) for C_{SF} gives

$$\frac{R_{PD}}{k_{PD}(A_P/V_P)} + C_{AS} = C_{SF} \qquad (2.5)$$

Substituting C_{AS} from Eq. (2.4) into Eq. (2.5) yields C_{SF}, thus

$$\frac{R_{Rxn}}{k_{Rxn}} + \frac{R_{PD}}{k_{PD}(A_P/V_P)} + C_{Eq} = C_{SF} \qquad (2.6)$$

Solving Eq. (2.3) for C_{SF} gives

$$\frac{R_{SFD}}{k_{SFD}(S_{Film}/V_{Film})} + C_{BF} = C_{SF} \qquad (2.7)$$

then substituting C_{SF} from Eq. (2.7) into Eq. (2.6) yields

$$\frac{R_{Rxn}}{k_{Rxn}} + \frac{R_{PD}}{k_{PD}(A_P/V_P)} + \frac{R_{SFD}}{k_{SFD}(S_{Film}/V_{Film})} + C_{Eq} = C_{BF} \qquad (2.8)$$

Rearranging Eq. (2.8) expresses the rate of reactant conversion in terms of C_{BF} and C_{Eq}. Thus

$$\frac{R_{Rxn}}{k_{Rxn}} + \frac{R_{PD}}{k_{PD}(A_P/V_P)} + \frac{R_{SFD}}{k_{SFD}(S_{Film}/V_{Film})} = C_{BF} - C_{Eq}$$

By assuming $R_{Rxn} = R_{PD} = R_{SFD} = R$, then rearranging the above equation, the overall rate of reactant conversion in terms of C_{BF} and C_{Eq} becomes

$$R\left\{\frac{1}{k_{Rxn}} + \frac{1}{k_{PD}(A_P/V_P)} + \frac{1}{k_{SFD}(S_{Film}/V_{Film})}\right\} = C_{BF} - C_{Eq} \qquad (2.9)$$

or

$$R = \left\{\frac{1}{(1/k_{Rxn}) + (1/k_{PD}(A_P/V_P)) + (1/k_{SFD}(S_{Film}/V_{Film}))}\right\} C_{BF} - C_{Eq}$$

The overall reaction rate constant is by definition

$$k_{Overall} = \left\{\frac{1}{(1/k_{Rxn}) + (1/k_{PD}(A_P/V_P)) + (1/k_{SFD}(S_{Film}/V_{Film}))}\right\}$$

Inverting k_{Overall} yields

$$\frac{1}{k_{\text{Overall}}} = \frac{1}{k_{\text{Rxn}}} + \frac{1}{k_{\text{PD}}(A_{\text{P}}/V_{\text{P}})} + \frac{1}{k_{\text{SFD}}(S_{\text{Film}}/V_{\text{Film}})} \qquad (2.10)$$

We can simplify Eq. (2.10) by combining $1/k_{\text{Rxn}}$ and $1/k_{\text{PD}}(A_{\text{P}}/V_{\text{P}})$. Doing so yields

$$\frac{1}{k_{\text{Overall}}} = \frac{k_{\text{PD}}(A_{\text{P}}/V_{\text{P}}) + k_{\text{Rxn}}}{k_{\text{PD}}(A_{\text{P}}/V_{\text{P}})k_{\text{Rxn}}} + \frac{1}{k_{\text{SFD}}(S_{\text{Film}}/V_{\text{Film}})} \qquad (2.11)$$

which reduces to

$$\frac{1}{k_{\text{Overall}}} = \frac{1}{\eta k_{\text{Rxn}}} + \frac{1}{k_{\text{SFD}}(S_{\text{Film}}/V_{\text{Film}})} \qquad (2.12)$$

where η is defined as

$$\eta = \frac{k_{\text{PD}}(A_{\text{P}}/V_{\text{P}})}{k_{\text{PD}}(A_{\text{P}}/V_{\text{P}}) + k_{\text{Rxn}}}$$

We call η the "effectiveness factor." The effectiveness factor accounts for the concentration difference along the pore of a solid-supported catalyst. If $k_{\text{PD}}(A_{\text{P}}/V_{\text{P}}) \gg k_{\text{Rxn}}$, i.e., if the fixed-bed reactor is reaction rate limited, then $\eta = 1$ and if $k_{\text{Rxn}} \gg k_{\text{PD}}(A_{\text{P}}/V_{\text{P}})$, i.e., if the fixed-bed reactor is pore diffusion rate limited, then $\eta < 1$. η depends only upon the pore structure of the solid-supported catalyst and is readily calculated via a variety of published methods.[3-7]

We can now specify R_{A} and substitute it into the component balance

$$\int_{C_{A,\text{In}}}^{C_{A,\text{Out}}} \frac{dC_{\text{A}}}{R_{\text{A}}} = \int_{0}^{V} \frac{dV_{\text{Fluid}}}{Q} = \frac{V_{\text{Fluid}}}{Q} \qquad (2.13)$$

and solve for C_{A}.

2.2 IMPACT OF DIFFUSION

Since the publication of Hougen and Watson's[8] classic volume *Chemical Process Principles: Kinetics and Catalysis* in 1947, the parameter used for characterizing flow through processes containing a solid, whether catalytic or noncatalytic, has been space velocity. Space velocity is Q/V, the inverse of space time V/Q. Note that we have dropped the subscript

"Fluid" on V; henceforth, V means the fluid volume of the catalyst mass, unless otherwise identified. Thus, space velocity has units of

$$\frac{(m^3/s)}{m^3} = \frac{1}{s}$$

Q/V has the same units as a first-order reaction rate constant. This equivalence between an inverse physical time and an inverse chemical reaction time is generally made when discussing a fixed-bed process. However, these two inverse times are only equivalent when the process is reaction rate limited. These two inverse times are not equivalent if the process is diffusion rate limited.

For a diffusion rate limited process, the rate of reactant consumption is

$$R_{\text{Diffusion}} = -k_{\text{Overall}}(C_{\text{BF}} - C_{\text{Eq}}) \tag{2.14}$$

and for a reaction rate limited process, the rate of reactant consumption is

$$R_{\text{Reaction}} = -k_{\text{Rxn}}(C_{\text{BF}} - C_{\text{Eq}}) \tag{2.15}$$

Both processes are described by a first-order equation; however, note that k_{Rxn} has dimensions of inverse chemical reaction time, while k_{Overall} has dimensions of inverse physical time. For a fixed-bed, catalyzed process that is reaction rate limited, we substitute Eq. (2.15) into Eq. (2.13) to get

$$\left(\frac{V}{Q}\right)_{\text{Reaction}} = \int_{C_{\text{BF,In}}}^{C_{\text{BF,Out}}} \frac{dC_{\text{BF}}}{R_{\text{BF}}} = \int_{C_{\text{BF,In}}}^{C_{\text{BF,Out}}} \frac{dC_{\text{BF}}}{k_{\text{Rxn}}(C_{\text{BF}} - C_{\text{Eq}})} \tag{2.16}$$

Rearranging, then integrating Eq. (2.16) yields

$$k_{\text{Rxn}}\left(\frac{V}{Q}\right)_{\text{Reaction}} = \ln\left[\frac{C_{\text{BF,Out}} - C_{\text{Eq}}}{C_{\text{BF,In}} - C_{\text{Eq}}}\right] \tag{2.17}$$

For a diffusion rate limited process, we substitute Eq. (2.14) into Eq. (2.13). Making these substitutions, then integrating yields

$$k_{\text{Overall}}\left(\frac{V}{Q}\right)_{\text{Diffusion}} = \ln\left[\frac{C_{\text{BF,Out}} - C_{\text{Eq}}}{C_{\text{BF,In}} - C_{\text{Eq}}}\right] \tag{2.18}$$

But, $(V/Q)_{\text{Reaction}}$ designates the time for a molecule of the bulk fluid, BF, to react while $(V/Q)_{\text{Diffusion}}$ designates the time for a molecule of the bulk fluid to diffuse, then react. Assuming that the right sides of Eqs. (2.17) and (2.18) are equal gives

$$
k_{Rxn} \left(\frac{V}{Q} \right)_{Reaction} = k_{Overall} \left(\frac{V}{Q} \right)_{Diffusion}
$$

or

$$
\frac{k_{Rxn}}{k_{Overall}} \left(\frac{V}{Q} \right)_{Reaction} = \left(\frac{V}{Q} \right)_{Diffusion}
$$

Substituting for $1/k_{Overall}$, then performing the necessary algebraic simplification yields

$$
k_{Rxn} \left\{ \frac{1}{k_{Rxn}} + \frac{1}{k_{PD} \left(A_P \Big/ V_P \right)} + \frac{1}{k_{SFD} \left(S_{Film} \Big/ V_{Film} \right)} \right\} \left(\frac{V}{Q} \right)_{Reaction}
$$

$$
= \left(\frac{V}{Q} \right)_{Diffusion}
$$

$$
\left\{ 1 + \frac{k_{Rxn}}{k_{PD} \left(A_P \Big/ V_P \right)} + \frac{k_{Rxn}}{k_{SFD} \left(S_{Film} \Big/ V_{Film} \right)} \right\} \left(\frac{V}{Q} \right)_{Reaction}
$$

$$
= \left(\frac{V}{Q} \right)_{Diffusion}
$$

$$
\tag{2.19}
$$

Thus, to obtain the same reactant conversion, the space time of a diffusion rate limited process must be greater than the space time for an equivalent reaction rate limited process. This point is important because, historically, we upscale and downscale fixed-bed processes using space velocity Q/V. When developing a new fixed-bed, catalyzed process, we upscale using laboratory and pilot plant reactor data to design commercial-sized reactors. When supporting an existing fixed-bed, catalyzed process, we downscale using commercial reactor data to design laboratory and pilot plant reactors. Table 2.1 contains the space velocities for various sized fixed-bed reactors for an actual process. The solid-supported catalyst is the same in each of the reactors listed in Table 2.1, as is the feed. The space velocities of Commercial—Plant 2, Catalyst Test Unit Pilot Plant, and Laboratory—Unit 2 are essentially the same: namely $(Q/V) = 10$. However, the linear, interstitial fluid velocities for these

Table 2.1 Comparison of Space Velocities for a Process Using Fixed-Bed Reactors			
Reactor Scale	Space Velocity (1/h)	Interstitial, Linear Fluid Velocity (m/h)	Reynolds Number
Commercial—Plant 1	18	165	92,500
Commercial—Plant 2	11	94	52,900
Recycle Pilot Plant	13	4	15
Catalyst Test Unit Pilot Plant	10	0.4	1
Laboratory—Unit 1	60	7	18
Laboratory—Unit 2	10	2	1

reactors are quite different. The Reynolds number for Commercial—Plant 2 is 52,900, while the Reynolds number for Laboratory—Unit 2 is one. The Commercial—Plant 2 reactor operates in the turbulent flow regime while the Laboratory—Unit 2 reactor operates in the laminar flow regime. These two reactors are comparable only if the process is reaction rate limited, which it is not. This process is diffusion rate limited; therefore, none of the laboratory or pilot plant reactors model the commercial plant reactors. If this point is not understood, then a great deal of money will be wasted doing experiments in the laboratory and pilot plant reactors, experiments that have no meaning for the commercial plant.

2.3 IMPLICATION OF FLOW REGIMES AND RATE CONTROLLING MECHANISMS

Commercial plants operate at high volumetric flow rates because their purpose is to produce as much product per unit time as possible. Laboratory and pilot plant facilities have a different purpose. Their purpose is to produce quality information that can be used to design a new commercial process or to support the operations of an existing commercial facility. We design commercial facilities to store large volumes of feed and product and to circulate large quantities of process fluids safely and with minimum environmental impact. We operate laboratory and pilot plant reactors at low volumetric flow rates because we want to minimize the volume of feed stored at the research facility. Another reason we operate laboratory and pilot plant reactors at low volumetric flow rate is waste disposal: the product produced by such reactors cannot be sold. Therefore, it must be disposed, which is expensive. Also, laboratory and pilot plant processes circulate small quantities of process fluids so that any leak is small, thereby minimizing the safety and environmental issues arising from the leak. Finally, if we are developing an entirely new process, we will be unsure of all the potential hazards it entails. Thus, we will keep all reactive volumes small to reduce the impact of a runaway or contaminated reaction. In other words, laboratory and pilot plant fixed-bed reactors generally operate in the laminar flow regime while commercial-sized fixed-bed reactors operate in the turbulent flow regime.

This flow regime difference can have a dramatic impact upon the results produced by a laboratory or pilot plant fixed-bed reactor. In the laminar flow regime, the stagnant film surrounding each solid-supported catalyst pellet or extrudate will be much thicker than the stagnant film surrounding solid-supported catalyst in a commercial reactor operating in the turbulent flow regime.[9] The diffusion rate constant, expressed as m/s, is the same in both flow regimes; however, the time to cross the stagnant film is larger for laminar flow than for turbulent flow since the stagnant film is thicker for laminar flow than for turbulent flow. Thus, product formation will be slower in the laminar flow regime than in the turbulent flow regime.

If the laboratory and pilot plant fixed-bed reactors are stagnant film diffusion rate limited, then $k_{Overall}$ will plot as scatter around an average value, which may or may not be recognized as k_{SFD}, the stagnant film diffusion rate constant. Not being aware of this potentiality produces highly expensive, inconclusive process support efforts and catalyst development programs. Years can be spent evaluating different catalyst sizes, shapes, and compositions to no avail because all the data scatters about one value. Unbeknownst to those working on the project, that point is the stagnant film diffusion rate constant. Eventually, the project will be canceled due to no conclusive results.

A similar situation can arise for pore diffusion rate limited processes. In this situation, the physical structure of the solid support must be altered to improve catalyst performance, i.e., to increase $k_{Overall}$. Changing the chemical composition of the solid-supported catalyst will not alter $k_{Overall}$. All the data produced by the laboratory and pilot plant reactors will scatter about one $k_{Overall}$, which is the pore diffusion rate constant for the process.

If a solid-supported catalyst is reaction rate limited, then changing size, shape, or pore structure will not improve catalyst performance. In this case, the only way to improve catalyst performance is to alter the chemical composition of the catalyst.

The message is: know the flow regime occurring in a fixed-bed reactor and know the rate controlling step of a solid-supported catalyzed process before attempting to improve process or catalyst performance.

2.4 IDENTIFICATION OF RATE CONTROLLING REGIMES

Consider Eq. (2.12), which is

$$\frac{1}{k_{Overall}} = \frac{1}{\eta k_{Rxn}} + \frac{1}{k_{SFD}(S_{Film}/V_{Film})} \qquad (2.12)$$

Equation (2.12) is analogous to the resistance present in an electrical circuit. In other words, the overall resistance is the sum of the individual resistances present in the circuit. In our case, we are measuring the resistance to form product molecules. We can determine $k_{Overall}$ directly by measuring $C_{BF,In}$ and $C_{BF,Out}$, then calculating $k_{Overall}$ from the below equation

$$R = k_{Overall}(C_{BF,Out} - C_{BF,In})$$

We can determine the value of k_{Rxn} from laboratory experiments that obviate any diffusion effects. What we do not know is $k_{SFD}(S_{Film}/V_{Film})$. However, we do know that k_{SFD} is constant at a given reactor operating temperature and pressure. Thus, only S_{Film}/V_{Film} responds to changes in the velocity of the bulk fluid over the surface of the catalyst. Remember S_{Film}/V_{Film} is the ratio of the stagnant film's surface area to its volume; therefore, it is inversely proportional to the stagnant film's thickness. Mathematically

$$\frac{S_{Film}}{V_{Film}} \propto \frac{1}{\delta}$$

where δ is the thickness of the stagnant film surrounding each solid-supported catalyst pellet or extrudate. But, stagnant film thickness is proportional to the Reynolds number of the bulk fluid, i.e.,[10]

$$\delta \propto \frac{1}{\sqrt{Re_z}} = \frac{1}{\sqrt{\rho D v_z/\mu}} = \sqrt{\mu/\rho D v_z}$$

Thus, as the interstitial, linear velocity of the bulk fluid increases, δ decreases. And, as δ decreases, reactant and product molecules spend less time traversing the stagnant film surrounding each solid-supported catalyst pellet or extrudate, which increases $k_{Overall}$. We can relate δ to v_z since the fluid density, fluid viscosity, and reactor diameter remain constant during the velocity change. Thus

$$\frac{S_{Film}}{V_{Film}} \propto f(v_z)$$

$f(v_z)$ represents an unspecified function of v_z. Equation (2.12) can now be written as

$$\frac{1}{k_{\text{Overall}}} = \frac{1}{\eta k_{\text{Rxn}}} + \frac{1}{k_{\text{SFD}} * f(v_z)}$$

which has the form of a straight line if we plot $1/k_{\text{Overall}}$ as a function of $1/f(v_z)$. The slope is $1/k_{\text{SFD}}$ and the intercept is $1/\eta k_{\text{PD}}$.

Figure 2.2 shows a schematic of such a plot. Drawing a horizontal line through the intercept demarcates film diffusion from pore diffusion and reaction rate. Below that horizontal line, the catalytic process is limited by a combination of pore diffusion rate and reaction rate.

Figure 2.3 shows actual data for olefin isomerization using a porous, solid acid catalyst. Figure 2.3 presents $1/k_{\text{Overall}}$ as a function of $1/\sqrt{v_z}$. It represents four fixed-bed laboratory and pilot plant reactors of different sizes. The data presented in Figure 2.3 was collected over a period of years. Considering the timeline of the project, $R^2 = 0.8553$ is quite good.

Figure 2.4 presents data collected over a number of years for olefin metathesis using a metal impregnated, solid-supported catalyst. The data is plotted as $1/k_{\text{Overall}}$ as a function of $1/\sqrt{v_z}$. The correlation is reasonable considering the time span of the project. Using plots similar to Figures 2.3 and 2.4, we can determine when a process shifts from

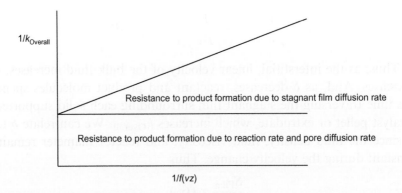

Figure 2.2 $1/k_{Overall}$ as a function of $1/f(v_z)$.

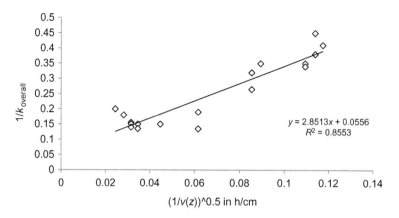

Figure 2.3 Olefin isomerization by porous, solid acid.

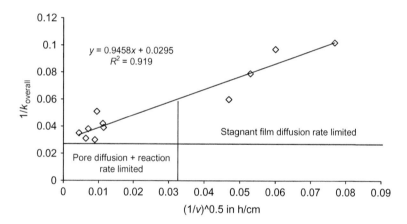

Figure 2.4 $1/k_{Overall}$ as a function of $(1/v)^{\wedge}0.5$ for Olefin metathesis.

being pore diffusion and reaction rate limited to being stagnant film diffusion rate limited. The linear correlation for Figure 2.4 is

$$\frac{1}{k_{Overall}} = 0.946\left(\frac{1}{\sqrt{v_z}}\right) + 0.0295$$

At $\left(1/\sqrt{v_z}\right) = 0.031$, $(1/k_{Overall}) = 0.0588$. Subtracting 0.0295 from 0.0588 gives 0.0293. At this value of $1/\sqrt{v_z}$, the resistance to product formation is balanced between pore diffusion plus reaction rate resistance and stagnant film diffusion resistance. In other words, the length from the horizontal axis to the horizontal line intersecting the y-axis equals the length of the line from intersecting horizontal line to the line representing the function. At $\left(1/\sqrt{v_z}\right) = 0.032$, the process is stagnant film diffusion rate limited; at $\left(1/\sqrt{v_z}\right) = 0.030$, the process is pore diffusion plus reaction rate limited.

We must know which resistance controls product formation before launching a catalyst or process improvement project. If stagnant film diffusion rate controls product formation, then the only way to increase product formation is to increase the interstitial fluid velocity through the catalyst mass. If pore diffusion plus reaction rate controls product formation, then we must first decide which resistance is the major component of their sum. Measuring $k_{Overall}$ for the solid-supported catalyst and for crushed catalyst is the most expeditious way to determine the presence of pore diffusion. If

$$k_{Overall}^{Pellet} = k_{Overall}^{Crushed}$$

then pore diffusion is not present during product formation. If

$$k_{Overall}^{Pellet} < k_{Overall}^{Crushed}$$

then product formation is pore diffusion rate limited. $k_{Overall}^{Crushed}$ is greater than $k_{Overall}^{Pellet}$ because the radius of the crushed pellet or extrudate is smaller than the radius of the whole pellet or extrudate, which means reactant and product molecules spend less time traversing the pores of the crushed sample compared to the whole sample. This result arises because k_{PD} is constant at controlled experimental temperatures and pressures. Unfortunately, this simple experimental test does not provide information about the extent of pore diffusion resistance.

To determine the extent of pore diffusion resistance, consider Eq. (2.10), which is

$$\frac{1}{k_{Overall}} = \frac{1}{k_{Rxn}} + \frac{1}{k_{PD}(A_P/V_P)} + \frac{1}{k_{SFD}(S_{Film}/V_{Film})}$$

However, when

$$\frac{1}{k_{Rxn}} + \frac{1}{k_{PD}(A_P/V_P)} \gg \frac{1}{k_{SFD}(S_{Film}/V_{Film})}$$

then the above resistance equation thus reduces to

$$\frac{1}{k_{\text{Overall}}} = \frac{1}{k_{\text{Rxn}}} + \frac{1}{k_{\text{PD}}(A_\text{P}/V_\text{P})}$$

In this equation, k_{Rxn} is a constant determined from laboratory experiments that obviated any diffusion resistance. k_{PD} is also a constant at a given operating temperature and pressure. Thus, to change k_{Overall}, we must change A_P/V_P. Catalyst manufacturers can change A_P/V_P for a given catalyst; however, A_P and V_P are reported as distributions and, while their averages for two solid-supported catalysts may be different, their distributions will most likely overlap, thereby making it difficult to interpret any changes in k_{Overall}.

However, A_P/V_P for catalyst extrudate is

$$\frac{A_\text{P}}{V_\text{P}} = \frac{\pi R_\text{P}^2}{\pi R_\text{P}^2 L_\text{P}} = \frac{1}{L_\text{P}}$$

where L_P is the average pore length. For a catalyst extrudate, L_P can run from the extrudate centerline to the surface, in which case it is equivalent to the radius of the extrudate. Or, L_P can run the length of the catalyst extrudate. The former situation implies that

$$\frac{A_\text{P}}{V_\text{P}} \propto \frac{S_{\text{Solid}}}{V_{\text{Solid}}}$$

Thus, we can replace A_P/V_P with $S_{\text{Solid}}/V_{\text{Solid}}$. Note that $S_{\text{Solid}}/V_{\text{Solid}}$ is the ratio of external surface area to volume for the solid support of the catalyst. Also, note that $S_{\text{Solid}}/V_{\text{Solid}}$ ratios change with solid-support size and shape. $S_{\text{Solid}}/V_{\text{Solid}}$ is the inverse of the solid's radius. Thus, as $S_{\text{Solid}}/V_{\text{Solid}}$ increases, the radius of the solid decreases. Hence, the time required for reactant and product molecules to traverse the pore network inside the solid-supported catalyst is reduced since the pore diffusion rate constant, k_{PD}, remains constant as the radius changes. Result: product formation increases with increasing $S_{\text{Solid}}/V_{\text{Solid}}$ if the process is pore diffusion rate limited.

Substituting $S_{\text{Solid}}/V_{\text{Solid}}$ for A_P/V_P in the above equation yields

$$\frac{1}{k_{\text{Overall}}} = \frac{1}{k_{\text{Rxn}}} + \frac{1}{k_{\text{PD}}(S_{\text{Solid}}/V_{\text{Solid}})}$$

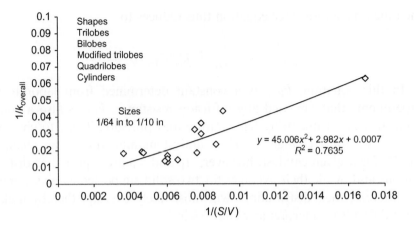

Figure 2.5 1/k$_{Overall}$ as a function of 1/(S/V) for Olefin isomerization by a solid-supported catalyst.

This equation suggests a linear relationship between $1/k_{Overall}$ and $1/(S_{Solid}/V_{Solid})$. However, A_P/V_P relates to S_{Solid}/V_{Solid} in some unspecified manner. We should, therefore, not expect a linear relationship between $1/k_{Overall}$ and $1/(S_{Solid}/V_{Solid})$. Under such a condition, we should use the best correlation relating $1/k_{Overall}$ to $1/(S_{Solid}/V_{Solid})$. Figure 2.5 shows a plot of $1/k_{Overall}$ as a function of $1/(S_{Solid}/V_{Solid})$ for olefin isomerization using a solid-supported catalyst. Stagnant film diffusion is negligible for these experiments. A polynomial provides the best fit for these data. As before, we can identify the $1/(S_{Solid}/V_{Solid})$ ratio where the resistance to product formation moves from reaction rate limited to pore diffusion rate limited. In Figure 2.5, at $1/(S_{Solid}/V_{Solid})$ of 0.000235, $1/k_{Overall}$ is 0.0014. Subtracting 0.0007 from 0.0014 leaves 0.0007. In other words, the resistance to reaction equals the resistance to pore diffusion. Thus, at values of $1/(S_{Solid}/V_{Solid}) < 0.000235$, the process is reaction rate limited. At all higher values, the process is pore diffusion rate limited.

2.5 IMPLICATIONS OF CONTROLLING RATE REGIMES

Pilot plant fixed-bed reactors are traditionally designed at space velocities equivalent to commercial scale fixed-bed reactors. Thus, the film diffusion resistance of the process at the two scales is different. In general, pilot plant fixed-bed reactors are film diffusion rate limited while commercial-size fixed-bed reactors are either pore diffusion rate or, more rarely, reaction rate limited. This shift from film diffusion rate limited to, most generally, pore diffusion rate limited occurs due to the high volumetric fluid flow through the catalyst mass in a commercial-size fixed-bed reactor. Thus, reactant consumption or product formation is faster in the commercial-size fixed-bed reactor than in the pilot plant fixed-bed reactor.

The same shift occurs when downsizing a process from the commercial scale to the pilot plant scale. If we use a space velocity equivalent to that of the commercial fixed-bed reactor to design a pilot plant, then the controlling resistance to reactant consumption or product formation shifts from pore diffusion or reaction rate to stagnant film diffusion rate. This shift adversely impacts the results of the research program for which the pilot plant was built. Downsized pilot plants are built to solve process problems or to develop new catalysts. If the process problems are related to stagnant film diffusion, then such a pilot plant will be useful; however, if the process problems are not related to stagnant film diffusion, then the pilot plant may produce spurious and misleading information.

With regard to developing new catalysts, if, at the commercial scale, the controlling resistance to reactant consumption or product formation is pore diffusion, then the median pore diameter (MPD), pore size distribution (PSD), and S_{Solid}/V_{Solid} ratio for the solid support must be optimized. However, if, at the pilot plant scale, the predominant resistance to reactant consumption or product formation is stagnant film diffusion, then changing solid support MPD and PSD will yield inconclusive results. Changing S_{Solid}/V_{Solid} will yield improved catalyst performance in the pilot plant but, depending on the extent of film diffusion resistance at the commercial scale, no improvement may be observed when the new catalyst is adopted as the catalyst of choice.

If, at the commercial scale, the predominant resistance to reactant consumption or product formation is reaction rate, then the chemistry of the catalyst must be optimized. However, the results from such an optimization program will be inconclusive if reactant consumption or product formation in the pilot plant is controlled predominately by stagnant film diffusion. Such a project may continue for a number of years before the inevitable conclusion is made: that the catalyst cannot be improved. Unfortunately, this conclusion will be drawn because the catalysts were evaluated in the wrong flow regime.

2.6 TEMPERATURE DEPENDENCE OF CONTROLLING REGIMES

All the resistances identified above are temperature dependent. At low temperature, the reaction rate is such that no concentration difference develops across the stagnant film or along the pores. In this case, intrinsic surface kinetics dominates the chemical process. We can visualize the active sites as being saturated with reactant. When the active sites are reactant saturated, a plot of $\ln(k_{\text{Overall}})$ versus $1/K$, where K is temperature in kelvins, is linear and its slope is $-E_{\text{Rxn}}/R$.

Increasing the process temperature increases the intrinsic reaction rate. At some temperature, a concentration difference develops along the pores. In other words, some active sites remain saturated with reactant while other active sites become reactant "starved." In this case, a plot of $\ln(k_{\text{Overall}})$ versus $1/K$ is a plot of $\ln(k_{\text{Rxn}} + k_{\text{PD}}(A_{\text{P}}/V_{\text{P}}))$ versus $1/K$. The slope of such a plot gives the apparent activation energy for reaction and pore diffusion. It is the arithmetic average of the energy of activation for the reaction and the energy of activation for pore diffusion. Thus, the slope is $-(E_{\text{Rxn}} + E_{\text{PD}})/2R$. If we crush the solid-supported catalyst, then measure k_{Overall} as a function of temperature, the new $\ln(k_{\text{Overall}}^{\text{Crushed}})$ plot will lie above the plot for $\ln(k_{\text{Overall}}^{\text{Pellet}})$. However, their slopes will be the same.[11]

As process temperature increases, the reaction rate becomes so high that only those active sites on or near the catalyst's surface "see" reactant. In other words, the pores of the solid-supported catalyst are filled with product molecules. The reactant concentration difference along the pore disappears. The only reactant concentration difference now occurs across the stagnant film surrounding each catalyst pellet or extrudate. Thus, the process is stagnant film diffusion rate limited. A plot of $\ln(k_{\text{overall}})$ versus $1/K$ is actually a plot of $\ln(k_{\text{SFD}}(S_{\text{Film}}/V_{\text{Film}}))$. The slope of such a plot is the activation energy for the diffusion of reactant in the bulk fluid, i.e., the slope is $-E_{\text{BD}}/R$, where BD identifies "bulk diffusion." For gases, this E_{BD} is $4-12$ kJ/mole; for liquid hydrocarbon, E_{BD} is $10-20$ kJ/mole; and, for aqueous solutions, E_{BD} is $8-10$ kJ/mole.[11(p320)]

Determining the apparent activation energy for a solid-supported catalytic process provides another method for identifying its controlling resistance, which must be known in order to design equipment and experimental programs.

2.7 DESIGN OF FIXED-BED REACTORS

Successfully scaling fixed-bed reactors requires using a $k_{Overall}$ for reactant consumption or product formation that reflects the appropriate process resistance at the design flow rate. Estimating such a global rate constant is best done from Figure 2.2, which is a generalization of Figures 2.3 and 2.4. Using such a figure to estimate $k_{Overall}$ at design interstitial fluid velocity insures using the appropriate resistance.

To generate Figure 2.2 requires several fixed-bed reactor sizes. Interstitial fluid velocities capable of minimizing stagnant film diffusion resistance must be used to approach the interstitial fluid velocities generally achieved through commercial scale fixed-bed reactors. The interstitial fluid velocities obtainable in laboratory scale fixed-bed reactors determine that portion of the relationship dominated by stagnant film diffusion resistance. Thus, it is important to have available fixed-bed reactors capable of operating at intermediate interstitial fluid velocities to establish the relationship between $1/k_{Overall}$ and $1/\sqrt{v_z}$. Once such a relationship is established, then $1/k_{Overall}$ can be determined for any design $1/\sqrt{v_z}$. $1/k_{Overall}$ can then be used in a process simulation to determine reactant consumption or product formation, thereby yielding information as to the financial viability of the project.

2.8 SUMMARY

This chapter analyzed the performance of fixed-bed reactors and discussed the importance of identifying the resistance controlling the rate of reactant consumption or product formation. This chapter stressed that the controlling resistance must be identified before designing a fixed-bed reactor or prior to planning an experimental program to support a fixed-bed, catalyzed process. This chapter also discussed the impact of diffusion on the overall rate constant for a catalyzed process. It also discussed the impact of temperature on the overall rate constant.

REFERENCES

1. Carberry J. *Can J Chem Eng* 1958;**36**:207.

2. Carberry J. *Chemical and catalytic reaction engineering*. New York, NY: McGraw-Hill, Inc.; 1976. p. 530–1.

3. Smith J. *Chemical engineering kinetics*. 3rd ed. New York, NY: Mc-Graw Hill Book Company; 1983. p. 477.

4. Satterfield C. *Mass transfer in heterogeneous catalysis*. Cambridge, MA: MIT Press; 1970. p. 129.

5. Satterfield C. *Heterogeneous catalysis in practice*. New York, NY: Mc-Graw Hill Book Company; 1980. p. 381.

6. Lee H. *Heterogeneous reactor design*. Boston, MA: Butterworth Publishers; 1985 [chapter 4].

7. Aris R. *The mathematical theory of diffusion and reaction in permeable catalysts*, vol. 1. Oxford, UK: Clarendon Press; 1975.

8. Hougen O, Watson K. *Chemical process principles: kinetics and catalysis*. New York, NY: John Wiley & Sons; 1947.

9. Hughes R. *Deactivation of catalysts*. London, UK: Academic Press; 1984. p. 2.

10. Granger R. *Fluid mechanics*. New York, NY: Dover Publications, Inc; 1995. p. 708 [First published by Holt, Reinhart and Winston, 1985].

11. Satterfield C. *Heterogeneous catalysis in practice*. New York, NY: Mc-Graw Hill Book Company; 1980. p. 321.

CHAPTER 3

Catalyst Deactivation

3.1 INTRODUCTION

Fixed-bed reactors, over time, show declining efficacy. This trend is independent of reactant feed concentration and reactor or process temperature.[1] It is due to declining catalyst productivity. In other words, as time progresses, the quantity of feed converted to product by the catalyst declines. At some point, the catalyst produces so little product that the chemical process becomes uneconomic. At that point, the catalyst bed or catalyst charge is "spent." We must either regenerate the spent catalyst *in situ* or we must dump the spent catalyst and charge the reactor with new catalyst or regenerate the spent catalyst "off-site." Regenerating spent catalyst off-site is not economic, except for catalysts containing exotic metals or precious metals.

The rate at which catalyst productivity declines varies with catalyst type and process. For example, petroleum cracking catalysts can become uneconomic within seconds of encountering hot, vapor phase hydrocarbon. However, catalysts used for ammonia synthesis demonstrate high productivity for a year or more. Naphtha reforming catalysts also demonstrate "service" lives of a year or more.[2]

Knowing the standard or accepted service life of a solid-supported catalyst allows us to determine the economics of a given process or plant. If that economic analysis produces a favorable result, then we will build the plant. In an economic analysis, we are little concerned with why the catalyst decays or how it decays. We are only concerned about the fact that it does decay. However, if, during its operation, the catalyst decay rate increases or market conditions for feed or product change, then we must know why and how our solid-supported catalyst decays so we can restore its service life and the economic viability of the process.

3.2 SOLID-SUPPORTED CATALYST DEACTIVATION

Solid-supported catalysts deactivate via two general mechanisms. One mechanism is spatially dependent, while the other mechanism is temporally dependent. By spatially dependent deactivation, we mean a reasonably sharp zone demarks active catalyst from inactive catalyst and that the zone moves through the catalyst bed as time passes. In other words, the volume of active catalyst becomes smaller as time passes. Thus, the amount of product produced per pass through the catalyst bed declines. Temporally dependent deactivation implies that catalyst performance declines as a function of "stream time" or operating time. In this mechanism, catalyst deactivation occurs randomly throughout the catalyst bed. Thus, for temporally dependent deactivation, there is no sharp zone demarking active catalyst from inactive catalyst.

Understanding either of these deactivation mechanisms requires the concept of "active site." By active site, we mean the location on the solid support where reactant undergoes "transformation" to product. Identifying the active site as a location raises several questions, the answers to which define the active site concept. Those questions are:[1(pp4−5)]

- What is the chemical composition of the active site?
- What chemical properties does the active site exhibit?
- What reactions occur at the active site?
- How does the structure of the support influence the nature of the active site?
- How many active sites per unit area exist on the solid support?
- How much area does a given solid support have?

Answering these questions not only defines the active site for a given catalyst, but they go far toward establishing the mechanism by which the catalyst deactivates.

3.3 CATALYST DEACTIVATION BY CHEMICAL MECHANISMS

Most fixed-bed reactors utilize one of three solid supports, even though the list of potential solid supports is quite large. Those three supports are:

1. metallic, pure metal or metal alloys;
2. metal oxide, such as alumina, silica, or magnesia;
3. mixed metal oxide, such as silica−alumina.

Common metal supports are nickel, used to catalyze various hydrogenation reactions, and iron, used to synthesize ammonia. The fact that a metal support acts as a catalyst for a particular chemical reaction implies that the metal atoms on its surface are not identical: some small number of metal atoms catalyze the chemical reaction, while the majority of metal atoms are inert toward the reactant. Since the mid-1970s, a cottage industry, whose goal is to identify these "special" metal atoms, has arisen and many papers have been published hypothesizing the nature of these special atoms. From our viewpoint, it is sufficient to admit their existence and accept their presence on the surface of a metal. From this admission, we infer that the catalytically active metal atoms have unoccupied d-orbitals since they are on the metal's surface. These unoccupied d-orbitals can interact with the electrons of a particular reactant molecule, thereby forming product, but these d-orbitals can also interact with the electrons of a molecule other than the reactant. We identify these other molecules as "poisons" if this interaction is strong enough to preclude further participation of the d-orbitals in the desired reaction. Thus, molecules possessing lone electron pairs, such as molecules containing VA and VIA elements, and molecules containing electron-rich multiple bonds, such as π-bonds, are potential poisons of active sites on metallic supports. Potential poisons possessing lone electron pairs are NH_3, as well as pyridines and pyrroles, PH_3 and other phosphines, AsH_3, H_2O, and organic alcohols, H_2S, and organic thiols and organic sulfides. Potential poisons possessing π-bonds are carbon monoxide, dienes, acetylenes, and aromatics.[1(p125)]

Metal oxides comprise the second group of catalyst supports. Interestingly, metal oxides, such as alumina and magnesia, respond to acid/base titration, thereby suggesting they behave chemically as acids or bases. Since metal oxides demonstrate acid and base characteristics, we can use them as solid acid or base catalysts. We can also react those acid and base sites with metals, thereby using the metal oxide as a solid support.

Acids involve either the transfer of a proton or the acceptance of an electron pair. A Bronsted acid donates a proton, H^+, to a molecule. The reaction of ammonia and hydrochloric acid

$$NH_3 + HCl \rightarrow NH_4^+ + Cl^-$$

is an example of a Bronsted acid donating a proton to a strong base. Bronsted acids also donate protons to weak bases. For example, strong

acids can protonate double bonds, as shown by the reaction of 1-butene with hydrochloric acid

$$H_2C=CHCH_2CH_3 + HCl \rightarrow H_3C - C^+HCH_2CH_3 + Cl^-$$

Lewis[3] proposed a more generalized definition for acids and bases. A Lewis acid accepts an electron pair from a molecule; conversely, Lewis bases donate electron pairs to Lewis acids. For example, a Lewis acid will accept a hydride ion, H^-, from a hydrocarbon molecule. In general, titrations do not distinguish between Bronsted acidity and Lewis acidity. Titrations report total acidity.

Alumina, Al_2O_3, the most common metal oxide support, has no significant Bronsted acidity at any calcination temperature. However, alumina does demonstrate Lewis acidity after high-temperature calcination. During high-temperature calcination, the alumina surface undergoes dehydration, thereby producing a Lewis acid site and a basic site, as shown below.

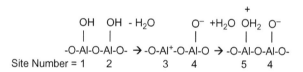

Site 3 above is a Lewis acid and site 4 above is a Lewis base. If we expose calcined alumina to atmospheric moisture, it will hydrate, thereby converting a Lewis acid site to a Bronsted acid site, as site 5 illustrates. Little work has been done to characterize the Lewis base sites on alumina.

Any molecule with an unshared electron pair can neutralize an acid site, thereby deactivating an active site. For example, three coordinate nitrogen has an unshared electron pair; therefore, it behaves as a Lewis base. Three coordinate phosphorus also has an unshared electron pair; thus, it behaves as a Lewis base. Also, any molecule containing a π-bond, i.e., a double bond or triple bond, behaves as a Lewis base. Thus, ammonia, amines, water, oxygenated hydrocarbons, phosphines, and alkenes and aromatics deactivate solid acid catalysts.

In a production facility, it is difficult to keep reactor feeds pristine. For example, all production facilities employ heat exchangers with water or steam as the cooling or heating fluid, respectively. Thus, water can leak into chemical processes, particularly if the pressure of the service fluid is greater than the pressure of the process fluid. Water ingress also occurs during feed and product storage. Most stored hydrocarbons accumulate 20–30 parts per million (ppm) water. Such absorbed water deactivates solid acid catalysts; thus, it must be removed from reactor feed prior to entering the fixed-bed reactor.

Air ingress into a hydrocarbon process forms oxygenated hydrocarbons. Pump seal leaks and flange leaks admit air, i.e., oxygen, into a hydrocarbon process. However, most air enters a hydrocarbon process during vacuum distillation. Ingressing oxygen reacts with hot hydrocarbon to form hydroperoxides, which then decompose by various mechanisms to alcohols and aldehydes or ketones. The resulting alcohols, aldehydes, and ketones behave as Lewis bases when they encounter a solid acid catalyst, thereby deactivating the catalyst.

Three coordinate nitrogen and phosphorus, i.e., amines and phosphines, possess an unshared electron pair. Thus, they behave as Lewis bases when they encounter a Lewis acid. Plant treated water contains amines and phosphines. They enter a chemical process through leaks of treated water into the process. Also, feed "as received" may contain trace quantities of amines and phosphines.

Alkene and aromatic π-bonds are electron-rich. Thus, a π-bond behaves as a Lewis base when it encounters a Lewis acid. In a double-bond isomerization process catalyzed by a solid acid, some fraction of the solid acid sites undergo an addition reaction with an alkene molecule, thereby forming an "addition complex." When such a complex forms, the acid site ceases to be catalytic. Such catalyst deactivation constitutes feed poisoning or "natural poisoning," which results in "natural catalyst deactivation."

Coke formation occurs at the acidic sites on alumina. These acidic sites create carbonium ions from feed molecules and/or product molecules, which react to form higher molecular weight entities. These higher molecular weight entities grow over the surface of the solid-supported catalyst as well as cyclize and lose hydrogen until they exhibit the properties of aromatics. As they grow, they not only cover solid support surface, but they also cover active sites; therefore, catalyst productivity decreases. Eventually, catalyst productivity becomes so poor that the catalyst must be regenerated. Coke formation can occur when feed encounters the first catalyst layer, then grow successively as a "front" through the catalyst bed. If coke forms via a reaction intermediate or the reaction product, it deposits randomly throughout the catalyst, accumulating as a function of "stream time."

3.4 CATALYST DEACTIVATION BY THERMAL MECHANISMS

Not all the surface atoms of a metallic support catalyze a chemical reaction. Only a small number of surface atoms on a metallic support interact with molecules in a second phase. These active surface metal atoms exist at a unique, geometric feature on the surface of the metallic support, such as a crystal defect. It is well known that crystal defects migrate and agglomerate or coalesce and disappear at crystal boundaries. When crystal defects agglomerate or coalesce, their d-orbitals rearrange and they become inactive with regard to chemical reactant. We call the movement and agglomeration of metal atoms "sintering." Sintering has been investigated intensely for many years, resulting in myriad proposed mechanisms. Suffice it to say: sintering happens. And, sintering is primarily temperature dependent: as temperature increases, the sintering rate increases. Sintering also depends on the surrounding environment: oxidizing environments accelerate sintering; reducing environments decelerate sintering.

Sintering begins at 600°C. At 600°C and above, metallic active sites diffuse across the metal's surface until they encounter another active site, at which time they coalesce to form agglomerations which are inactive toward reactant molecules. Thus, at 600°C and above, metal-supported catalysts demonstrate declining productivity due to sintering. Such sintering is more rapid in oxidizing milieus than in reducing milieus. Some metal-supported catalysts demonstrate active site redispersion between 500°C and 600°C in oxidizing atmospheres; however, such active site redispersion is uncommon and requires experimental proof per catalyst.[1(chp 5)]

The crystalline phase of the active site may also be thermally unstable. High temperatures during process excursions and combustion regeneration may induce a phase change at the active site, from a catalytically active crystallinity to a catalytically inactive crystallinity. Such a change has an adverse impact on catalyst productivity.[4]

The metal oxides, including those used as catalyst solid supports, are refractories. They are not melted and poured into molds or pulled into wire or compressed into given shapes. Their forms and structure come from mixing metal oxide powder with water to create a "mull," which has a consistency of bread dough. This mull is then pumped by an extruder through a die plate. Extrusion through a die plate forms metal oxide cylinders, bilobes, trilobes, or quadrilobes of various "diameters." Metal oxide spheres or pellets are formed in tilted, rotating drums. The moist metal oxide shape is then dried to remove water and calcined to set the pore surface area and porosity of the final product.

Metal oxide shapes have two defining structures: a macrostructure and a microstructure. Their macrostructure reflects their origin: they are hydrated powder, thus they are powder particles adhering to each other. The space between powder particles forms macropores. However, each powder particle was formed via crystallization of the metal oxide from a "mother liquor." As such, each powder particle is comprised of myriad adhering crystals. Metal oxides are thus polycrystalline. The space between two such adhering crystals forms a micropore. Macropores are low pore surface area structures while micropores are high pore surface area structures.

Sintering during calcination sets the pore surface area and porosity of the shaped metal oxides. In this case, sintering does not mean the diffusion of metal atoms and their agglomeration. With regard to metal oxides, sintering means the densification of the polycrystalline metal oxide shape. Take a loose sheet of paper and fold it in half. Imagine each facing page to be a crystalline metal oxide surface and imagine the space between these surfaces to be a metal oxide micropore. Each crystalline surface represents an energy expenditure by Nature. It is the energy required to retain the surface atoms in the parent crystal. As the surrounding temperature increases, the juncture of the two crystalline planes moves toward the pore mouth, thereby releasing energy through surface reduction. The result is densification since pore volume disappears and is replaced by solid. Thus, the average pore size of the metal oxide shape increases due to the disappearance of micropores.[5,6] And, as average pore size increases and surface area decreases, the number of active sites declines, as does catalyst productivity.

The surface area and average pore size for metal oxide pellets or extrudates remain constant so long as they are not subjected to temperatures near or above those used during their manufacture. Metal oxides undergo sintering at temperatures near their manufacturing temperature and they undergo rapid sintering at temperatures above their manufacturing temperature. Coke removal by combustion regeneration reaches these temperatures.

Nearly all the solid-supported catalysts used by the CPI are solid acids that induce the formation of carbonaceous material, i.e., coke, that accumulates on the metal oxide's surface. Eventually, the performance of the solid-supported catalyst becomes so low that we isolate the effected fixed-bed reactor from the process and either dump the coked catalyst and fill the reactor with a fresh catalyst charge or we combustion regenerate the coked catalyst in the fixed-bed reactor. The latter course of action is the most common one. Thus, a solid-supported catalyst within a fixed-bed reactor will undergo a number of combustion regenerations, each of which induces sintering of the metal oxide support during the high-temperature period of the regeneration procedure. Due to this sintering, the solid-supported catalyst loses some number of active sites, which causes a decline in catalyst productivity with each regeneration. This decline in catalyst activity from one to another operating campaign may be small and hardly noticeable, but over a long time period, it becomes substantial.

Figure 3.1 shows the idealized performance for an olefin metathesis catalyst. Such catalysts are molybdenum, tungsten, or rhenium on alumina. Catalyst productivity starts at 100% for a fresh charge of

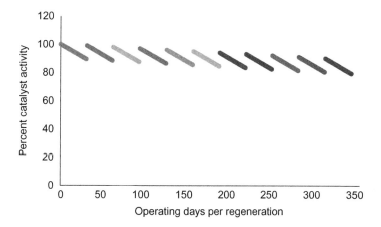

Figure 3.1 Performance of olefin metathesis catalyst as a function of operating days and regeneration.

catalyst and during a thirty day operating campaign falls to 90% due to coke formation. At this point, we isolate the fixed-bed reactor and combustion regenerate the solid-supported catalyst. Following the first regeneration, the fixed-bed reactor resumes production at 99% catalyst productivity; after 30 days of operation, catalyst productivity is 89%. Again, we isolate the fixed-bed reactor and combustion regenerate the solid-supported catalyst. Figure 3.1 shows 10 production campaigns following catalyst regeneration. The 1% drop in catalyst productivity following a regeneration might not be noticeable by the operating staff, but when fixed-bed reactor performance is displayed as in Figure 3.1, the performance trend is clear. Each regeneration induces a step change in catalyst activity. At some point in time, the operating staff will decide to dump the current catalyst charge and replace it with a new catalyst charge.

3.5 CATALYST DEACTIVATION BY MECHANICAL MECHANISMS

Crush and attrition are the most common mechanical mechanisms causing poor catalyst productivity. Solid-supported catalyst must be able to support its own weight after being charged to a reactor. The catalyst at the bottom of the reactor will be crushed if a solid-supported catalyst cannot support its own weight.

Crushing of solid-supported catalyst produces fines which cause fluid channeling through the catalyst mass or completely block fluid flow through the catalyst mass. Channeling causes the formation of flow "tubes" through the catalyst mass in a fixed-bed reactor. The high fluid velocity along and through these tubes yields a short residence time for reaction, which reduces catalyst productivity. Also, the catalyst in such tubes undergoes wave front poisoning and/or coking at an accelerated rate since most of the fresh feed flows through them.

Channeling induces fluid stagnation in the catalyst mass. Fluid wets the catalyst in regions of stagnation, thus reaction occurs. However, product from these regions contributes little to catalyst performance since the fluid does not exit them or exits them slowly. We can consider stagnant regions of fixed-bed reactors as low flow continuous reactors or as batch reactors, both with poor agitation. These stagnant regions can develop into "hot spots" within the catalyst mass where temperatures become high enough to induce crystalline phase changes at active sites, thereby causing catalyst deactivation. Hot spots also demonstrate accelerated coke formation, which provides another mechanism for reducing catalyst productivity within the stagnant flow region.

Attrition occurs when catalyst pellets or extrudates rub each other, thereby creating fines. Attrition occurs when high-velocity gases impinge the catalyst mass, thereby causing pellet or extrudate fluidization. Such fluidization generally occurs during combustion regeneration. The regenerating gas enters the fixed-bed reactor through a nozzle; thus, it enters at high velocity. If the top of the catalyst mass is close to the nozzle, the impinging high-velocity gas excavates a pit in it. The excavated pellets or extrudates are fluidized, which induces attrition and fines formation. Fines also form at the point of breakage if breakage occurs during fluidization. Pellet or extrudate fluidization ceases once a deep pit is excavated into the catalyst mass. At that time, the fines settle into the catalyst mass, as they do when gas flow stops. Pressure drop across the fixed-bed reactor begins to increase as fines accumulate in the catalyst mass. With time, fines accumulation becomes great enough to induce fluid channeling through the catalyst mass. Fines accumulation may become great enough to completely block flow through the catalyst mass, thereby necessitating the catalyst's removal and replacement.

Chart 3.1 presents the various catalyst deactivation mechanisms in an organizational schematic. It provides a method for diagnosing a catalyst deactivation event at a production facility. Note that it is difficult to distinguish between spatially and temporally dependent catalyst deactivation for thermal and mechanical mechanisms.

3.6 SOLID BASES AND NEUTRAL SOLIDS

The surfaces of alkaline earth metal oxides, such as MgO, CaO, SrO, and BaO, demonstrate chemical basicity. The basic site on these alkaline earth metal oxides consists of strongly basic O^{2-} centers; little work has been done beyond identifying the strongly basic sites on alkaline earth metal oxides.[7] Bronsted acids and Lewis acids deactivate alkaline earth metal oxide catalysts. Alkaline earth metal oxides are also refractories; thus, they deactivate via sintering if heated to high temperatures. However, they do not coke.

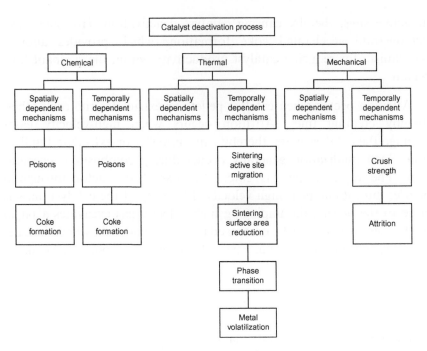

Chart 3.1 Basic mechanisms causing solid-supported catalyst deactivation. (Adapted with permission from J. H. Worstell, M. J. Doll, and J. M. R. Ginestra, "What's Causing Your Catalyst to Decay?", Chemical Engineering Progress, September 2000, pages 59–64.)

Silica, SiO_2, possesses neither Bronsted nor Lewis acidity. Lewis acidity is not generated in SiO_2 by high-temperature calcination. Pure silica is neutral—as we would expect glass to be.[7] Silica is used as a catalyst solid support, but it is difficult to bind metals to it.

3.7 SILICA–ALUMINA OXIDE CATALYSTS

We do mix silica with alumina to increase the acidity of a given metal oxide support. Silica–alumina mixtures have both Bronsted acidity and Lewis acidity. In silica–alumina, the ratio of Bronsted acid sites to Lewis acid sites depends upon calcination temperature since Bronsted acid sites are converted quantitatively into Lewis acid sites upon heating.

Some controversy exists as to the structure of solid acid sites on silica–alumina. When a trivalent aluminum atom is substituted for a quadravalent silicon atom in a silica matrix, a net negative charge develops on the aluminum atom. This net negative charge is stabilized by a proton, which comes from the dissociation of water. Thus, substituting trivalent aluminum for quadravalent silicon produces a Bronsted acid site.

The acidic properties of silica–alumina depend upon

• preparation method;
• proportion of alumina to silica;
• dehydration temperature;
• hydration, i.e., steam treatment.

The maximum number of Bronsted acid sites in silica–alumina occurs near 25 weight percent alumina. This weight percent alumina in silica corresponds to uniformly distributed alumina moieties in a silica matrix. At higher weight percent alumina in silica, the alumina moieties begin to interconnect and form an alumina matrix. Silica–alumina generally displays maximum acidity and catalytic activity at Al/Si atomic ratios less than one, again indicating the importance of the Al–O–Si linkages in the silica matrix. Thus, the manufacturing procedure used to produce a given silica–alumina solid support significantly impacts the performance of a particular catalyst.

Heating silica–alumina drives off "water of constitution," thereby converting Bronsted acid sites into Lewis acid sites. The primary acid formed in silica–alumina is most likely a Lewis acid site, which subsequently converts to a Bronsted acid site upon contact with humid air. Thus, Bronsted acid site formation is a second-order process.[8]

3.8 BIFUNCTIONAL SOLID-SUPPORTED CATALYSTS

Many solid-supported catalysts used in the CPI are bifunctional, meaning they perform two or more reactions simultaneously. Most such catalysts are metal oxides onto which a metal has been deposited or grafted. In most cases, bifunctional catalysts are alumina or silica–alumina that has been pore impregnated with a solution containing the appropriate metal. They are then drained and calcined. Bifunctional catalysts can also be manufactured by co-mulling alumina or silica–alumina with a transition metal oxide.

Petroleum is reformed using bifunctional catalysts. These catalysts are generally platinum or palladium on alumina or silica–alumina.

The acid sites of the alumina or silica—alumina isomerize the petroleum molecules, which are gasoline-grade molecules, and the platinum or palladium dehydrogenates and hydrogenates the gasoline molecules. If the reforming process is reaction rate limited, then the porous alumina is saturated with H_2PtCl_6. Following pore saturation, the solid support is drained and calcined in air, which oxidizes the platinum or palladium to an oxide. Exposing the metal oxide—platinum oxide or —palladium oxide mixture to hydrogen reduces the platinum or palladium to a zero valent state. The platinum or palladium occurs as 8—100 Angstrom crystallites on the alumina or silica—alumina surface. In other words, each crystallite possesses a number of unoccupied d-orbitals.[9]

If the reforming process is diffusion rate limited, then only the outer portion of the porous metal oxide is pore impregnated. The remainder of the catalyst production process is similar to that for a pore saturated catalyst. This manufacturing procedure produces a partial pore impregnation that forms a "shell" of reactive platinum crystallites on or near the exterior surface of the metal oxide support.

Hydrodesulfurization of catalytic reforming feeds is another example of bifunctional catalysis. Hydrodesulfurization catalysts remove organic sulfur from petroleum feed stocks. These catalysts possess hydrogenation capacity, although not nearly the hydrogenation capacity of a reforming catalyst. All such processes inject hydrogen into the petroleum feed prior to its entering the fixed-bed reactor. The hydrogen interacts with the hydrodesulfurization catalyst to produce H_2S, NH_3, paraffins, and metals. The metals deposit on the surface of the mixed oxide support, while the other products are solubilized by the petroleum solvent.[10]

Hydrodesulfurization catalysts are alumina that have been comulled with molybdenum oxide and/or cobalt oxide, thereby forming a porous mixed metal oxide support. This mixed metal oxide support can be pore saturated with either a moybdenate or cobalt solution to bring the catalyst's metal content to 10—20%. The impregnated mixed metal oxide catalyst then undergoes calcination in air, thereby producing cobalt molybdate, which is an industry term, not a scientific term.[11] The cobalt molybdate, oxide catalyst is then treated with H_2S to sulfide its surfaces prior to entering operational service. Historically, the oxide catalyst was sulfided in the petroleum process. Today, many such catalysts are sulfided, then delivered to the refinery for immediate use.

Since hydrodesulfurization catalysts contain solid acid sites and transition metal crystallites, they are susceptible to the same poisons as metal oxide catalysts and metal-supported catalysts, namely, Lewis bases and π-bonds in organic molecules. However, the major mechanisms causing hydrodesulfurization catalyst deactivation are coking and deposition of metal sulfides. Initially, coke accumulates rapidly on the catalyst's surface, then coke deposition assumes a constant rate.[12] The rate of coke deposition depends upon the hydrogen partial pressure within the fixed-bed reactor: low hydrogen partial pressure yields a high coke deposition rate while high hydrogen partial pressure reduces coke deposition.[9(chp 1)] Coke deposition occurs uniformly throughout each hydrodesulfurization catalyst extrudate.

However, the metal sulfides formed during hydrodesulfurization do not distribute evenly throughout each catalyst extrudate; rather, they deposit in the peripheral regions of each catalyst extrudate. Those metal sulfides depositing in pore entrances restrict diffusion of petroleum molecules into the catalyst's interior. These metal sulfide deposits eventually grow large enough to block pore entrances. Such deposits deactivate a hydrodesulfurization catalyst permanently, whereas coke deposits can be removed by combustion regeneration, thereby restoring catalyst productivity to the extent that petroleum molecules can access the regenerated regions of each catalyst extrudate.

Some bifunctional solid-supported catalysts experience active site volatilization at temperatures inducing sintering. Examples of active site volatilization include[9(p18)]:

- the Deacon process which employs a supported copper catalyst that forms volatile $CuCl_2$ under operating conditions;
- the ruthenium-based process for NO_x reduction—unfortunately, under oxidizing atmospheres at high-temperature, supported ruthenium forms volatile RuO_3-RuO_4 moieties (this process was abandoned due to such volatilization);
- the Wacker oxidation process—incorporating the catalytic metal in a support makes the metal volatile under operating conditions.

Active site volatilization eventually requires catalyst replacement.

3.9 REACTION SCHEMES FOR CATALYST DEACTIVATION

Poisoning occurs via the chemical interaction of a particular type of molecule with the catalytically active site. In most cases, it is difficult to establish the exact reaction mechanism that occurs at an active site, but by proposing simple generalized reaction schemes, we can achieve significant understanding of what happens during poisoning of a solid-supported catalyst in a fixed-bed reactor. The same point is true for coke formation on a solid-supported catalyst.

Consider a fixed-bed reactor containing a solid-supported catalyst with active sites S. These active sites convert reactant R in the feed into product P. However, the feed to the fixed-bed reactor also contains a poison Poi that interacts irreversibly with active sites, thereby deactivating them. We can devise a simple reaction scheme for what occurs within the fixed-bed reactor: when reactant R encounters an active S, it interacts with S to form a complex R−S, which undergoes reaction to a complex P−S that subsequently dissociates into product P and active site S. S is now available to repeat the process. A second, independent reaction also occurs within the same fixed-bed reactor. When poison Poi encounters an active site S, it interacts with S to form a complex Poi−S. If the interaction between Poi and S is strong, the complex Poi−S is unlikely to dissociate, in which case, the active site S becomes permanently unavailable to reactant R. Thus, the active site is permanently poisoned. Reaction Scheme 3.1 portrays this sequence of events occurring within a fixed-bed reactor.

Even though Reaction Scheme 3.1 is simple, it provides insight into the poisoning of the catalyst mass. For example, if the interaction between Poi and S is rapid, then Poi will interact with the first S it encounters and forms the inactive complex Poi−S. Thus, a layer of inactive catalyst will form near the feed nozzle of the fixed-bed reactor. As time proceeds, this layer of inactive catalyst grows progressively through the catalyst mass until it reaches the discharge nozzle of the fixed-bed reactor. In other words, a definite, moving boundary forms

$$R + S \rightarrow R\text{-}S \rightarrow P\text{-}S \rightarrow P + S$$

$$Poi + S \rightarrow Poi\text{-}S$$

Reaction Scheme 3.1

between pristine catalyst and adulterated catalyst. We call this sequence of poisoning events "moving front" or "wave front" poisoning. Wave front poisoning does not depend upon operating time, i.e., on how long the fixed-bed reactor operates, rather wave front poisoning depends upon the quantity of feed entering the fixed-bed reactor since the feed contains the poison. Therefore, in wave front poisoning, declining catalyst productivity, i.e., catalyst deactivation, is spatially dependent.

But, what if the formation of Poi–S is not rapid? If the interaction between Poi and S is not strong and rapid, then Poi upon encountering S may simply diffuse past it. Such bypassing of S may occur many times before Poi finally binds irreversibly with an S. In this case, no definite boundary forms between pristine catalyst and adulterated catalyst. Thus, as operating time for the fixed-bed reactor increases, ever more active sites become inactive and catalyst productivity declines as a result of poisoning. We call this sequence of poisoning events "stream-time" poisoning. Stream-time poisoning does not depend upon the quantity of feed entering the fixed-bed reactor, rather it depends upon the time Poi remains in the fixed-bed reactor. In other words, given enough time, Poi will eventually find an S to deactivate.

In some processes, the reactant R is actually the poison that deactivates the catalyst mass. Reaction Scheme 3.2 portrays this sequence of events.

This reaction scheme contains three cases: two limiting cases and a case intermediate between the limiting cases. The first limiting case is that the first reaction is rapid and predominates. The second limiting case is that the second reaction is rapid and predominates. The intermediate case hypothesizes that neither reaction is fast relative to the other nor does either reaction predominate.

If the first reaction of Reaction Scheme 3.2 is rapid and all R quantitatively converts to P upon contact with the catalyst mass, then the remainder of the catalyst mass "sees" only P. Thus, no poisoning of S

$$R + S \rightarrow R\text{-}S \rightarrow P\text{-}S \rightarrow P + S$$

$$R + S \rightarrow R\text{-}S \rightarrow Poi\text{-}S$$

Reaction Scheme 3.2

occurs in it. In this case, wave front poisoning occurs. If the second reaction of Reaction Scheme 3.2 is rapid and predominates, then no P forms and the catalyst mass only "sees" R. Again, wave front poisoning occurs. In this case, no P appears in the fluid exiting the fixed-bed reactor. For the intermediate case, where neither reaction is rapid nor predominant relative to the other, both R and P stream through the catalyst mass. In this case, catalyst deactivation occurs via stream-time poisoning.

Product poisoning is yet another possibility for deactivating solid-supported catalyst in a fixed-bed reactor. Reaction Scheme 3.3 portrays this not uncommon potentiality.

If the dissociation of P–S is slow compared to the formation of Poi–S, then catalyst deactivation will be spatially dependent. If P–S dissociates faster than Poi–S forms, then catalyst deactivation will be temporally dependent.

Coke formation may be initiated by reactant, product, or both. Acid sites on solid supports initiate coke formation, as do catalytically active metal sites. Both sites initiate coke formation on bifunctional catalysts. Reaction Scheme 3.4 presents a simplified representation of coke formation.

Needless to say: coke formation is complex. Thus, it is best to keep to a simple mechanism to explain it. If the second and third reactions above are fast relative to the first and fourth reactions, then catalyst

$$R + S \rightarrow R\text{-}S \rightarrow P\text{-}S \rightarrow P + S$$

$$P + S \rightarrow P\text{-}S \rightarrow Poi\text{-}S$$

Reaction Scheme 3.3

$$R + S \rightarrow R\text{-}S \rightarrow P\text{-}S \rightarrow P + S$$

$$R + S \rightarrow R\text{-}S \rightarrow P\text{-}S \rightarrow Coke\text{-}S$$

$$R + S \rightarrow R\text{-}S \rightarrow Coke\text{-}S$$

$$P + S \rightarrow P\text{-}S \rightarrow Coke\text{-}S$$

Reaction Scheme 3.4

coking will resemble wave front poisoning. If the second and third reactions above are slow relative to the first and fourth reactions, then catalyst coking will resemble stream-time poisoning.

3.10 CATALYST DEACTIVATION PARAMETER

We need to know two parameters when operating a fixed-bed reactor. Those parameters are the overall rate constant $k_{Overall}$, discussed in Chapter 2, and the decay rate constant k_{Decay}. Note, we do not specify k_{Decay} for poisoning, coke formation, or sintering because we do not know which catalyst deactivation is under way within the fixed-bed reactor. In fact, they all may be adversely impacting catalyst productivity simultaneously. Thus, it is meaningless to propose and derive a detailed mechanism for catalyst deactivation since we do not know how many or which catalyst deactivation processes are occurring in our fixed-bed reactor. From an operating viewpoint, i.e., "pounds or kilograms out the door" viewpoint, we are not greatly interested in the actual values for $k_{Overall}$ or k_{Decay}. Our interest with regard to these parameters is whether they are changing or not. If they are changing, then the process in the fixed-bed reactor is not as it was. It requires investigation.

To determine that the process under way in our fixed-bed reactor is changing implies that we defined a "standard condition." In most cases, we define our standard condition to be a fresh catalyst charge and the product being produced at or near capacity. The component balance for A in our fixed-bed reactor is

$$\int_{C_{A,In}}^{C_{A,Out}} \frac{dC_A}{R_A} = \int_0^V \frac{dV_{Fluid}}{Q} = \frac{V_{Fluid}}{Q}$$

where C_A, V_{Fluid}, and Q were defined in Chapter 2. R_A is generally defined as

$$R_A(t = 0) = k_{Overall} N(t = 0) W C_A$$

where $N(t = 0)$ is the number of active sites per catalyst unit mass and W is the mass of catalyst in the fixed bed reactor. Note that N is time dependent. Therefore, after a given time period, R_A will be

$$R_A(t) = k_{Overall} N(t) W C_A$$

where $N(t)$ is the number of active sites per catalyst unit mass at time t. Dividing the above equation by the equation for $R_A(t=0)$ gives

$$\frac{R_A(t)}{R_A(t=0)} = \frac{k_{\text{Overall}}N(t)WC_A}{k_{\text{Overall}}N(t=0)WC_A} = \frac{N(t)}{N(t=0)}$$

assuming k_{Overall}, W, and C_A remain constant during the given time period. Rearranging the above equation yields

$$R_A(t) = \left(\frac{N(t)}{N(t=0)}\right) R_A(t=0)$$

Thus, we can obtain $R_A(t)$ for any time by multiplying $R_A(t=0)$ by the ratio of the number of active sites at time t divided by the number of active sites initially present in the catalyst mass. It is impossible to determine either $N(t=0)$ or $N(t)$; therefore, we determine the ratio itself. This ratio is "catalyst activity" and we designate it as "a"; thus

$$a = \left(\frac{N(t)}{N(t=0)}\right)$$

Levenspiel[13] presents a procedure for determining k_{Decay} for fixed-bed reactors. His procedure does not specify whether k_{Decay} is temporally or spatially dependent. Levenspiel assumes the rate at which catalyst activity changes with respect to time can be described as a power law equation, namely,

$$\frac{da}{dt} = -k_{\text{Decay}}a$$

Integrating the above equation yields

$$a_t = a_0\, e^{-k_{\text{Decay}}t}$$

where a_t is catalyst activity at any time greater than $t=0$. a_0 is catalyst activity at $t=0$ and is taken as unity.

The component balance for a first-order, irreversible chemical reaction in a fixed-bed reactor is

$$\int_{C_{A,\text{In}}}^{C_{A,\text{Out}}} \frac{dC_A}{R_A} = \int_{C_{A,\text{In}}}^{C_{A,\text{Out}}} \frac{dC_A}{-k_{\text{Overall}}C_A} = \frac{V_{\text{Fluid}}}{Q}$$

This mass balance assumes the catalyst never loses its functional activity. Incorporating the above activity function into the mass balance to account for loss of catalyst activity yields

$$\int_{C_{A,\text{In}}}^{C_{A,\text{Out}}} \frac{dC_A}{-k_{\text{Overall}} a_t C_A} = \frac{V_{\text{Fluid}}}{Q}$$

Substituting for a_t, then rearranging gives

$$\int_{C_{A,\text{In}}}^{C_{A,\text{Out}}} \frac{dC_A}{C_A} = -k_{\text{Overall}}(a_0 \, e^{-k_{\text{Decay}} t}) \frac{V_{\text{Fluid}}}{Q}$$

Integrating the above equation yields

$$\ln\left(\frac{C_{A,\text{Out}}}{C_{A,\text{In}}}\right) = -k_{\text{Overall}}(a_0 \, e^{-k_{\text{Decay}} t}) \frac{V_{\text{Fluid}}}{Q}$$

which gives, upon rearranging

$$-\ln\left(\frac{C_{A,\text{Out}}}{C_{A,\text{In}}}\right) = k_{\text{Overall}}(a_0 \, e^{-k_{\text{Decay}} t}) \frac{V_{\text{Fluid}}}{Q} = \ln\left(\frac{C_{A,\text{In}}}{C_{A,\text{Out}}}\right)$$

Taking the logarithm a second time yields

$$\ln\ln\left(\frac{C_{A,\text{In}}}{C_{A,\text{Out}}}\right) = \ln\left(a_0 k_{\text{Overall}} \frac{V_{\text{Fluid}}}{Q}\right) - k_{\text{Decay}} t$$

However, $a_0 = 1$, thus the above equation becomes

$$\ln\ln\left(\frac{C_{A,\text{In}}}{C_{A,\text{Out}}}\right) = \ln\left(k_{\text{Overall}} \frac{V_{\text{Fluid}}}{Q}\right) - k_{\text{Decay}} t$$

Therefore, plotting $\ln\ln(C_{A,\text{In}})/(C_{A,\text{Out}})$ as a function of time t produces a straight line with a slope equal to k_{Decay} and an intercept equal to $\ln(k_{\text{Overall}}(V_{\text{Fluid}}/Q))$ for a first-order, irreversible reaction. Thus, k_{Overall} can be calculated from the intercept of this plot.

We made three implicit assumptions in the above analysis of k_{Decay}. The first assumption is that reaction rate is separable from catalyst activity. In other words, at any given time, we assume the reaction rate is

$$R_{\text{Reaction}} = f(a) * g(C, T)$$

where $f(a)$ is a function describing catalyst activity at the specified time relative to a standard condition and $g(C,T)$ is a function describing the concentration and temperature dependence of the reaction rate. $g(C, T)$ is the kinetics of the reaction. Note that $g(C,T)$ is time independent while $f(a)$ is time dependent. We also assumed the catalyst deactivation rate is

$$R_{\text{Deactivation}} = \varphi(a) * \lambda(C, T)$$

where $\varphi(a)$ is a function describing the rate of change of catalyst activity and $\lambda(C,T)$ is a function describing the impact of concentration and temperature on catalyst deactivation. Thus, "separability" assumes that reaction kinetics does not change as catalyst productivity, i.e., catalyst activity, changes. In other words, reaction kinetics is independent of the number of active sites on a catalyst. Namely, R_{Reaction} is high if there are many active sites on the catalyst and R_{Reaction} is low if there are few active sites on the catalyst. Also, the shift from high R_{Reaction} to low R_{Reaction} is linear, which means $a = 1$ for fresh catalyst and $a = 0$ for completely deactivated catalyst.

All the active sites on the catalyst must have identical chemical and geometric characteristics for reaction kinetics to be independent of the number of active sites. That is, the active sites must be homogeneous, which is the second assumption we made in the above analysis of k_{Decay}.

The third assumption we made was that

$$a = \left(\frac{N(t)}{N(t = 0)} \right)$$

which is not necessarily true for all reactions. Catalyst activity may be the square or cube of $N(t)/N(t = 0)$.

We use this method, with all its assumptions, caveats, and restrictions, because it is simple and because it works. First, we generally do not know the reaction kinetics occurring at a solid-supported active site. We can hypothesize one or more active intermediates at the active site and we can draw structures and reaction paths for each, but whether they actually exist or not, we do not know. Second, we generally do not know how the geometric characteristics of the active site impact reaction kinetics. In light of these facts, we should make the simplest assumption that reaction kinetics can be separated from catalyst activity. Second, plotting $\ln\ln(C_{A,\text{In}}/C_{A,\text{Out}})$ versus time will confirm the validity of the second assumption. If the plot is linear, then we can say the active sites are homogeneous. If the plot is nonlinear, then we can assume the active sites are nonhomogeneous. While a linear plot does not prove our assumption, it does indicate the validity of our assumption. With regard to the third assumption, mechanisms can be hypothesized that yield

$$a = \left(\frac{N(t)}{N(t=0)} \right)^2$$

or

$$a = \left(\frac{N(t)}{N(t=0)} \right)^3$$

but such mechanisms are laden with assumptions and are so complex that they would be difficult to prove.

In summary, the above analysis of k_{Decay} rests upon three assumptions that cannot be proven; however, the analysis is simple and it does provide information about k_{Decay}, as Figure 3.2 shows. We simply need to remember that the plots in Figure 3.2 rest upon unprovable assumptions.

3.11 WAVE FRONT OR STREAM-TIME POISONING

After determining k_{Decay}, the question becomes: does catalyst deactivation occur via wave front poisoning or stream-time poisoning? Active

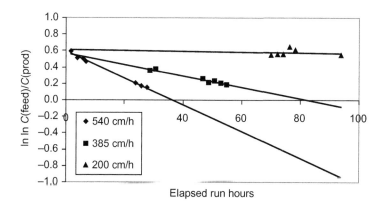

Figure 3.2 Isomerization as a function of elapsed run hours. Isomerizations performed at different superficial velocities. Laboratory sized fixed-bed reactors. (Reprinted with permission from J. H. Worstell, M. J. Doll, and J. M. R. Ginestra, "What's Causing Your Catalyst to Decay?", Chemical Engineering Progress, September 2000, pages 59–64.)

sites on a solid-supported catalyst become inactive when a poison P adsorbs irreversibly to them. The component balance for poison P passing through a catalyst mass is

$$\frac{\partial P}{\partial t} + \left(v_r \frac{\partial P}{\partial r} + \frac{v_\theta}{r} \frac{\partial P}{\partial \theta} + v_z \frac{\partial P}{\partial z} \right) = D_{AB} \left(\frac{1}{r} \frac{\partial}{\partial r} \left(r \frac{\partial P}{\partial r} \right) + \frac{1}{r^2} \frac{\partial^2 P}{\partial \theta^2} + \frac{\partial^2 P}{\partial z^2} \right) + R_P$$

If radial, axial, and azimuthal diffusion are negligible and if flow only occurs in the axial direction z, then the component balance for P becomes

$$\frac{\partial P}{\partial t} + v_z \frac{\partial P}{\partial z} = R_P$$

At steady state, the above equation reduces to

$$v_z \frac{\partial P}{\partial z} = R_P$$

where v_z is fluid velocity through the catalyst mass (m/s); z is the axial length of the catalyst mass (m); P is poison concentration in the fluid (mols $P/m^3{}^*$s); and R_P is the disappearance of P from the fluid (mols $P/m^3{}^*$s). The disappearance of P from the fluid occurs when P interacts with and adsorbs to the solid-supported catalyst. The amount of P adsorbed to the solid-supported catalyst is

$$\frac{W}{\rho_{CBD}} \left(\frac{1 - \varepsilon}{\varepsilon} \right) \frac{dn}{dt}$$

where W is the weight of the solid-supported catalyst in the fixed-bed reactor (kg); ρ_{CBD} is the compacted bulk density of the solid-supported catalyst (kg/m^3); ε is the fluid fraction and $1 - \varepsilon$ is the solid fraction of W; n is the moles of poison P adsorbed by the solid-supported catalyst (mols P/m^3 solid); and t is time (h). The units for the above equation are thus

$$\frac{W}{\rho_{CBD}} \left(\frac{1 - \varepsilon}{\varepsilon} \right) \frac{dn}{dt} = \left(\frac{\text{kg catalyst}}{\text{kg catalyst}/m^3} \right) \left(\frac{\text{solid volume fraction}}{\text{fluid volume fraction}} \right)$$

$$\times \left(\frac{\text{mols } P}{m^3 \text{ solid} * s} \right) = \frac{\text{mols } P}{m^3 \text{ fluid} * s}$$

We can therefore equate

$$\frac{W}{\rho_{CBD}}\left(\frac{1-\varepsilon}{\varepsilon}\right)\frac{dn}{dt} = R_P$$

and the component balance for P becomes

$$v_z\frac{dP}{dz} = -\frac{W}{\rho_{CBD}}\left(\frac{1-\varepsilon}{\varepsilon}\right)\frac{dn}{dt}$$

If we assume adsorption is stagnant film diffusion rate limited, then

$$\frac{W}{\rho_{CBD}}\left(\frac{1-\varepsilon}{\varepsilon}\right)\frac{dn}{dt} = k_{SFD}(S_{Solid}/V_{Solid})(P - P_{Sat})$$

where k_{SFD} is the film diffusion rate constant (m/s); S_{Solid}/V_{Solid} is the surface area to volume ratio for the catalyst pellet or extrudate (m^2/m^3); P is poison concentration in the fluid (mols P/m^3); and P_{Sat} is the poison saturation concentration (mols P/m^3). Substituting the above assumption into the component balance for P yields

$$v_z\frac{dP}{dZ} = -k_{SFD}(S_{Solid}/V_{Solid})(P - P_{Sat})$$

Rearranging and writing the equation as an integral gives

$$\int_{P_{Feed}}^{P}\frac{dP}{(P - P_{Sat})} = -\frac{k_{SFD}(S_{Solid}/V_{Solid})}{v_z}\int_0^Z dz$$

Performing the integration yields the equation for wave front adsorption passing through the catalyst mass.

$$\ln\left[\frac{P - P_{Sat}}{P_{Feed} - P_{Sat}}\right] = -\frac{k_{SFD}(S_{Solid}/V_{Solid})z}{v_z}$$

But, adsorption deactivates active sites on the solid-supported catalyst. Therefore, we can equate

$$k_{Decay} = k_{SFD}(S_{Solid}/V_{Solid})$$

and obtain

$$\ln\left[\frac{P - P_{Sat}}{P_{Feed} - P_{Sat}}\right] = -\frac{k_{Decay}z}{v_z}$$

which is equivalent to

$$\ln\left[\frac{P_{\text{Feed}} - P_{\text{Sat}}}{P - P_{\text{Sat}}}\right] = \frac{k_{\text{Decay}}z}{v_z}$$

Rearranging the above equation yields

$$\frac{v_z}{z}\ln\left[\frac{P_{\text{Feed}} - P_{\text{Sat}}}{P - P_{\text{Sat}}}\right] = k_{\text{Decay}}$$

Thus, k_{Decay} depends upon the fluid velocity through the catalyst mass, the concentration of poison in the feed, and the inverse length of catalyst mass. This relationship provides a diagnostic tool for differentiating between wave front and stream-time poisoning of solid-supported catalyst: if experiments show catalyst deactivation depends on any of these variables, then catalyst deactivation occurs by wave front poisoning; if catalyst productivity is independent of these variables, then catalyst deactivation occurs by stream-time poisoning. Figure 3.3 shows the impact of solid-supported catalyst bed length on k_{Decay}: as bed length increases, k_{Decay} decreases, as suggested by the above equation and shown by the slopes of the plots in Figure 3.3. Also, plotting k_{Decay} against fluid velocity through the catalyst mass produces a straight line, as shown in Figure 3.4. Such a plot can also demonstrate a shift in catalyst deactivation mechanism. In Figure 3.4, at 60°C and 80°C, this particular olefin isomerization catalyst deactivates by wave front poisoning. But the catalyst deactivation mechanism shifts to stream-time poisoning at 120°C. At 120°C, k_{Decay} is

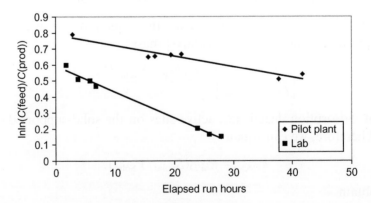

Figure 3.3 Isomerization as function of elapsed run hours. Pilot plant sized (5 in length) versus laboratory sized (2 in length) fixed-bed reactors. Superficial velocities of 500 cm/h. (Reprinted with permission from J. H. Worstell, M. J. Doll, and J. M. R. Ginestra, "What's Causing Your Catalyst to Decay?", Chemical Engineering Progress, September 2000, pages 59–64.)

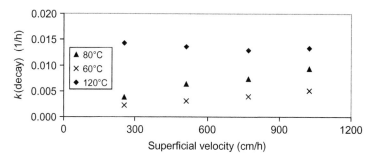

Figure 3.4 Isomerization catalyst decay rate constant as a function of superficial velocity through laboratory sized fixed-bed reactors. (Reprinted with permission from J. H. Worstell, M. J. Doll, and J. M. R. Ginestra, "What's Causing Your Catalyst to Decay?", Chemical Engineering Progress, September 2000, pages 59–64.)

independent of fluid velocity through the catalyst mass, which suggests that catalyst deactivation by a poison in the feed is no longer the dominant mechanism; instead, at 120°C, a different catalyst deactivation mechanism is dominant.

With regard to stream-time poisoning, reconsider the non-steady-state component balance for poison P. It is

$$\frac{\partial P}{\partial t} + v_z \frac{\partial P}{\partial z} = R_P$$

For stream-time poisoning, the removal of P from the fluid never reaches steady state, nor does it depend upon location z in the catalyst mass. Therefore, the above equation reduces to

$$\frac{\mathrm{d}P}{\mathrm{d}t} = R_P$$

But, from above

$$R_P = -k_{\text{SFD}}(S_{\text{Solid}}/V_{\text{Solid}})(P - P_{\text{Sat}}) = -k_{\text{Decay}}(P - P_{\text{Sat}})$$

Therefore,

$$\frac{\mathrm{d}P}{\mathrm{d}t} = -k_{\text{Decay}}(P - P_{\text{Sat}})$$

Rearranging and integrating the above equation gives

$$\int_{P_{\text{Feed}}}^{P} \frac{\mathrm{d}P}{(P - P_{\text{Sat}})} = -k_{\text{Decay}} \int_{0}^{t} \mathrm{d}t$$

Performing the integration yields

$$\ln\left[\frac{P - P_{Sat}}{P_{Feed} - P_{Sat}}\right] = -k_{Decay}t$$

which is equivalent to

$$\ln\left[\frac{P_{Feed} - P_{Sat}}{P - P_{Sat}}\right] = k_{Decay}t$$

or

$$\left(\frac{1}{t}\right)\ln\left[\frac{P_{Feed} - P_{Sat}}{P - P_{Sat}}\right] = k_{Decay}$$

Thus, catalyst deactivation via stream-time poisoning depends upon the concentration of poison in the feed and time. Stream-time poisoning does not depend upon fluid velocity through the catalyst mass nor does it depend upon the length of the catalyst mass.

To determine whether a catalyst deactivates via stream-time poisoning, start flow to the fixed-bed reactor and monitor the reaction by

$$\ln\ln\left(\frac{C_{A,In}}{C_{A,Out}}\right) = \ln\left(k_{Overall}\frac{V_{Fluid}}{Q}\right) - k_{Decay}t$$

or a similar equation. When the reaction is linear versus time, block flow to the fixed-bed reactor and let it sit at its operating temperature for a specified time, then resume flow to the reactor, as shown in Figure 3.5. Replot the data after removing time for quiescent operation. If the replotted data displays an offset, then stream-time poisoning is

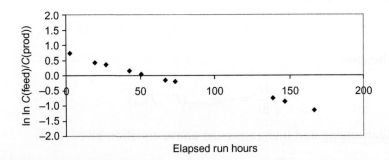

Figure 3.5 Isomerization as a function of elapsed run hours. Laboratory sized fixed-bed reactors. 120° C; superficial velocity 1080 cm/h. (Reprinted with permission from J. H. Worstell, M. J. Doll, and J. M. R. Ginestra, "What's Causing Your Catalyst to Decay?", Chemical Engineering Progress, September 2000, pages 59–64.)

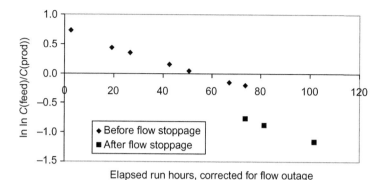

Figure 3.6 Isomerization as a function of elapsed run hours, corrected for flow outage. Laboratory sized fixed-bed reactors. 120° C; superficial velocity 1080 cm/h. (Reprinted with permission from J. H. Worstell, M. J. Doll, and J. M. R. Ginestra, "What's Causing Your Catalyst to Decay?", Chemical Engineering Progress, September 2000, pages 59–64.)

operative, as shown in Figure 3.6. If the replotted data rejoins and extends the original line, then wave front poisoning is operative.

In order to efficiently manage the reactor, we need to know whether the catalyst contained in a fixed-bed reactor deactivates by wave front poisoning or stream-time poison. If a given solid-supported catalyst deactivates via wave front poisoning, then we can fill it with process fluid at process temperature immediately after replacing spent catalyst with fresh catalyst or immediately after regenerating the catalyst *in situ*. Thus, the fixed-bed reactor is ready for immediate use when needed. Also, solid-supported catalysts that deactivate via wave front poisoning can be "blocked-in" during process upsets, then brought "online" after resolving the process upset.

If a given solid-supported catalyst deactivates via stream-time poisoning, which generally means feed molecules poison active sites or coke formation occurs on the catalyst's surface, then the solid-supported catalyst begins deactivating upon contact with process fluid, especially at process temperature. Filling such fixed-bed reactors prior to bringing them "online" is a mistake: while waiting to be streamed, they are undergoing catalyst deactivation. If we wait long enough, we may actually stream a fixed-bed reactor that has little or no catalyst productivity. Also, solid-supported catalysts that deactivate via stream-time poisoning should not be "blocked-in" during process upsets. If such fixed-bed reactors are "blocked-in" and it requires significant time to resolve the process upset, then, upon restarting the process, we may actually stream a completely deactivated catalyst mass, which raises a different set of upset issues. Therefore, fixed-bed reactors containing solid-supported catalyst that deactivates via stream-time poisoning should be drained during process upsets. If we know the upset to be of short duration, then we need to make a conscious decision about draining the reactor or not draining the reactor.

3.12 SELECTIVITY AND CATALYST DEACTIVATION

When managing the performance of a fixed-bed reactor, we are concerned with the amount of product produced by it relative to the amount of feed entering it and we are concerned about the amount of feed converted to an undesirable product relative to the desired product. The former concern is the "yield" of the fixed-bed reactor while the latter concern is the "selectivity" of the fixed-bed reactor.

Consider the reaction mechanism

$$A + B \rightarrow C$$
$$B + C \rightarrow P$$
$$A + C \rightarrow U$$

where A and B are reactants and B is limiting, i.e., A does not change concentration. C is an intermediate and P is the desired product and U is the undesired product. A, B, C, P, and U have units of mols/m^3. The rate equations for this reaction mechanism are

$$-\frac{dB}{dt} = k_1 \varphi(B)$$

$$\frac{dC}{dt} = k_1 \varphi(B) - k_2 \kappa(B)\gamma(C) - k_3 \eta(C)$$

$$\frac{dP}{dt} k_2 \kappa(B)\gamma(C)$$

$$\frac{dU}{dt} = k_3 \eta(C)$$

where k_1, k_2, and k_3 are rate constants and $\varphi(B)$, $\kappa(B)$, $\gamma(C)$, and $\eta(C)$ are functions describing the dependence of rate on the designated concentration. The differential yield of P is

$$\frac{dP/dt}{-(dB/dt)} = \frac{k_2 \kappa(B)\gamma(C)}{k_1 \kappa(B)} = \left(\frac{k_2}{k_1}\right)\gamma(C)$$

and the yield of P_U is

$$\frac{dU/dt}{-(dB/dt)} = \frac{k_3\eta(C)}{k_1\kappa(B)}$$

Therefore, the differential, point, or local selectivity of P_D to P_U is

$$\frac{dP/dt}{dU/dt} = \frac{k_2\kappa(B)\gamma(C)}{k_3\eta(C)} = \left(\frac{k_2}{k_3}\right)\frac{\kappa(B)\gamma(C)}{\eta(C)}$$

or

$$\frac{dP}{dU} = \left(\frac{k_2}{k_3}\right)\frac{\kappa(B)\gamma(C)}{\eta(C)}$$

The overall selectivity for this reaction mechanism is

$$\int_0^{P_D} dP = \left(\frac{k_2}{k_3}\right)\frac{\kappa(B)\gamma(C)}{\eta(C)}\int_0^{P_U} dU$$

which, upon integration and rearrangement, gives

$$\frac{P}{U} = \left(\frac{k_2}{k_3}\right)\frac{\kappa(B)\gamma(C)}{\eta(C)}$$

Now, assume the above reaction mechanism occurs on a bifunctional, solid-supported catalyst. Assume that reactions

$$A + B \rightarrow C$$
$$A + C \rightarrow U$$

occur on active site 2 of which there are N_2 sites and that reaction

$$B + C \rightarrow P$$

occurs on active site 1 of which there are N_1 sites. Modifying the equation for overall selectivity for this assumption gives us

$$\frac{P}{U} = \left(\frac{k_2}{k_1k_3}\right)\frac{N_1\gamma(C)}{N_2\eta(C)}$$

If a poison interacts preferentially with active site 1, then P/U decreases because N_1 decreases, i.c., overall selectivity decreases. If, however, a poison preferentially interacts with active site 2, then P/U increases because N_2 decreases, i.e., overall selectivity increases. Thus, shifts in overall selectivity indicate that poisoning is under way in a fixed-bed reactor containing a bifunctional, solid-supported catalyst.

3.13 SUMMARY

This chapter discussed solid-supported catalyst deactivation. It introduced the concept of catalytic active site and presented the various mechanisms causing active site deactivation. This chapter also defined and described spatially dependent catalyst deactivation and temporally dependent catalyst deactivation and provided methods for distinguishing between the two.

REFERENCES

1. Butt J, Petersen E. *Activation, deactivation, and poisoning of catalysts.* San Diego, CA: Academic Press; 1988. p. 1.

2. Hughes R. *Deactivation of catalysts.* London, UK: Academic Press; 1984, page 1.

3. Lewis bases have a pair of electrons available for donation to a Lewis acid. This electron pair can be unshared, as in the case of NH_3, or they can be in a π-bond, such as an olefin bond. A Lewis acid is any chemical species having a vacant orbital available for electron sharing. In a Lewis acid/base reaction, the unshared electron pair of the base forms a covalent bond with the vacant orbital of the acid.

4. Worstell J, Doll M, Worstell J. What's causing your catalyst to decay? *Chem Eng Progress* 2000.. p. 59.

5. West A. *Solid state chemistry and its applications.* Chichester, UK: John Wiley & Sons; 1984 [chapter 20].

6. Barsoum M. *Fundamentals of ceramics.* New York, NY: McGraw-Hill Companies, Inc.; 1997 [chapter 10].

7. Tanabe K. *Solid acids and bases: their catalytic properties.* New York, NY: Academic Press; 1970. p. 53.

8. Basila M, Kantner T, Rhee K. The Nature of the Acidic Sites on a Silica-Alurnina. Characterization by Infrared Spectroscopic Studies on Trirnethylamine and Pyridine Chemisorption. *J Phys Chem* 1964;**68**:3197.

9. Gates B, Katzer J, Schuit G. *Chemistry of catalytic processes.* New York, NY: McGraw-Hill Book Company; 1979. p. 237.

10. Gray J, Hardwerk G. *Petroleum refining: technology and economics.* New York, NY: Marcel Dekker, Inc; 1975. p. 114.

11. Bland W, Davidson R, editors. *Petroleum processing handbook.* New York, NY: McGraw-Hill Book Company; 1967. p. 3–38.

12. Mosby J, Hockstra G, Kleinhanz T, Sroka J. Pilot Plant Proves Resid Process. *Hydrocarbon Process* 1973;**52**(2):93–7.

13. Levenspiel O. Experimental Search for a Simple Rate Equation to Describe Deactivating Porous Catalyst Particles. *J Catal* 1972;**25**:265.

CHAPTER 4

Improving Fixed-Bed Reactor Performance

4.1 INTRODUCTION

We live at a time when communicating with others globally is, essentially, instantaneous. It is also a time when we can ship product cheaply, worldwide. Therefore, commercial competition exists in the extreme. The pressure to increase process productivity and efficiency, i.e., to reduce operating costs, is high. One component of this drive is improving fixed-bed reactor productivity and efficiency, which means improving solid-supported catalyst performance. In today's commercial environment, we must be committed to the continuous improvement of the solid-supported catalyst charged to our fixed-bed reactors.

Before initiating an effort to improve the productivity and efficiency of a solid-supported catalyst, we must ask:

1. Is reactant consumption or product formation diffusion rate or reaction rate limited?
2. Are by-products formed, and if so, what are their formation mechanisms?
3. Is deactivation spatially dependent or temporally dependent?

Development of a catalyst improvement program depends upon the answers to these questions.

4.2 IMPROVING PRODUCT FORMATION RATE

Consider the overall rate constant in its resistance form; it is

$$\frac{1}{k_{\text{Overall}}} = \frac{1}{k_{\text{Rxn}}} + \frac{1}{k_{\text{PD}}(A_{\text{P}}/V_{\text{P}})} + \frac{1}{k_{\text{SFD}}(S_{\text{Film}}/V_{\text{Film}})}$$

where k_{Rxn} is the reaction rate constant at the catalytic site (1/s), k_{PD} is the pore diffusion mass transfer rate constant (m/s), A_{P} is the average cross-sectional area of a pore (m^2), V_P is average pore volume (m^3), k_{SFD} is the stagnant film mass transfer rate constant (m/s), S_{Film} is the stagnant film surface area (m^2), and V_{Film} is the stagnant film volume (m^3) of the stagnant film surrounding the catalyst pellet or extrudate. The abbreviated form of the above equation is

$$\frac{1}{k_{Overall}} = \frac{1}{\eta k_{Rxn}} + \frac{1}{k_{SFD} * f(v_z)}$$

where η is the effectiveness factor for a given solid-supported catalyst and f(v) is a function describing linear fluid flow through the catalyst mass.

These two equations provide insight for improving the performance of a given fixed-bed reactor. If the fixed-bed reactor is stagnant film diffusion rate limited, then

$$\frac{1}{k_{SFD} * f(v_z)} \quad \text{or} \quad \frac{1}{k_{SFD}(S_{Film}/V_{Film})}$$

control the rate of reactant consumption or product formation. Therefore, increasing the linear fluid flow through the catalyst mass decreases the thickness of the stagnant film surrounding each catalyst pellet or extrudate, which increases the rate of reactant consumption or product formation. Alternatively, increasing the surface to volume ratio of the fluid surrounding each catalyst pellet or extrudate will increase reactant consumption or product formation. We generally equate S_{Film}/V_{Film} with S_{Solid}/V_{Solid} of the extrudate since we can quantify the latter. Any change in S_{Solid}/V_{Solid} produces an incremental change in S_{Film}/V_{Film}. From an operational viewpoint, we would first maximize the linear fluid flow through the catalyst mass, then we would optimize S_{Film}/V_{Film}, which means maximizing S_{Solid}/V_{Solid}. The maximum achievable S_{Solid}/V_{Solid} for a solid-supported catalyst depends upon the maximum allowable pressure drop Δp for a given fixed-bed reactor. The Δp across a catalyst mass depends upon the fluid's linear velocity, density, and viscosity, as well as the diameter of the solid support and the characteristic length of the solid support. We define the characteristic of the solid support as

$$L_{Char} = \frac{(W_{Solid}/\rho_{LBD})}{\varepsilon A_{CS}} = \frac{A_{CS} Z_{Solid}}{\varepsilon A_{CS}} = \frac{Z_{Solid}}{\varepsilon}$$

where W_{Solid} is the weight of solid-supported catalyst charged to the fixed-bed reactor (kg); ρ_{LBD} is the loose bulk density of the solid-supported catalyst (kg/m^3); A_{CS} is the cross-sectional area of the empty fixed-bed reactor

(m^2); and ε is the void fraction of the solid-supported catalyst. Z_{Solid} is the height of the catalyst mass in the reactor (we will use L_{Char} in Chapter 5). Note that εA_{CS} gives the flow area through the catalyst mass. Also, note that $Q/\varepsilon A_{CS}$ gives the linear fluid velocity through the catalyst mass when Q is volumetric fluid flow rate (m^3/s or h).

Increasing S_{Solid}/V_{Solid} actually decreases the radius and length of the solid-supported catalyst pellet or extrudate. For example, consider a cylindrical catalyst extrudate

$$\frac{S_{Solid}}{V_{Solid}} = \frac{2\pi r^2 + 2\pi r L}{\pi r^2 L}$$

$$\frac{S_{Solid}}{V_{Solid}} = \frac{\pi r(2r + 2L)}{\pi r(rL)}$$

$$\frac{S_{Solid}}{V_{Solid}} = \frac{2}{L} + \frac{2}{r}$$

where r is the radius of the solid support and L is the length of the solid support. Thus, the geometric size of the solid-supported catalyst decreases as S_{Solid}/V_{Solid} increases, which means it packs more efficiently in the reactor and its void fraction decreases, which causes a concomitant increase in Δp. Thus, for a given fixed-bed reactor, maximum allowable Δp sets the maximum possible S_{Solid}/V_{Solid}. Figure 4.1 shows this point schematically by plotting fixed-bed reactor Δp as a function of S_{Solid}/V_{Solid} and solid-supported catalyst shape. The shapes portrayed are cylinder, trilobe, bilobe, and quadrilobe. At a given Δp, the S_{Solid}/V_{Solid} increases as

$$(S_{Solid}/V_{Solid})_{Cylinder} < (S_{Solid}/V_{Solid})_{Trilobe}$$
$$< (S_{Solid}/V_{Solid})_{Bilobe} < (S_{Solid}/V_{Solid})_{Quadrilobe}$$

The measured overall rate constant increases similarly at the same Δp, namely

$$k_{Overall}^{Cylinder} < k_{Overall}^{Trilobe} < k_{Overall}^{Bilobe} < k_{Overall}^{Quadrilobe}$$

Thus, after maximizing the linear fluid flow through the catalyst mass for a given solid-supported catalyst shape, we then maximize S_{Solid}/V_{Solid} for that shape. If more catalyst productivity is required, then testing and qualifying a solid-supported catalyst shape that offers a greater S_{Solid}/V_{Solid} at the same Δp should be considered.

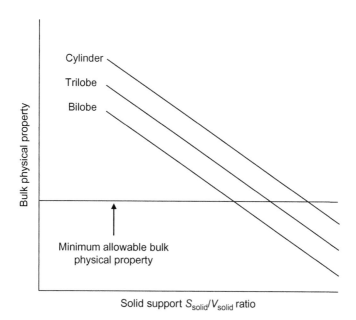

Figure 4.1 Bulk physical property (such as crush) as a function of solid support S_{Solid}/V_{Solid}. (Adapted with permission from J. H. Worstell and J. H. Worstell, "Improve Fixed-bed Reactor Performance without Capital Expenditure", Chemical Engineering Progress, January 2004, pages 51–57.)

The crush strength of the solid-supported catalyst may preclude reaching the maximum allowable pressure drop of a given fixed-bed reactor. Crush strength, in general, decreases as S_{Solid}/V_{Solid} increases. Also, crush strength decreases as the number of lobes on the solid support increases. When a solid support breaks, either across its radius, thereby forming two smaller pellets or extrudates, or along its axis, thereby forming two cylinders from a bilobe extrudate or one cylinder and a bilobe from a trilobe extrudate, it generates fines along the fragment line. These fines migrate through the catalyst mass until they lose "lift," at which point they "settle" and begin to accumulate. With time, the Δp of the fixed-bed reactor begins to rise. Eventually Δp reaches the maximum allowable Δp for the fixed-bed reactor, at which time the reactor must be isolated and the fines physically removed from the catalyst mass or the catalyst mass must be dumped and new catalyst charged to the reactor. Figure 4.2 shows these trends schematically.

Resistance to attrition increases as S_{Solid}/V_{Solid} increases. Thus, the solid support generates fewer fines by frictional abrasion, i.e., rubbing against each other, as S_{Solid}/V_{Solid} increases.

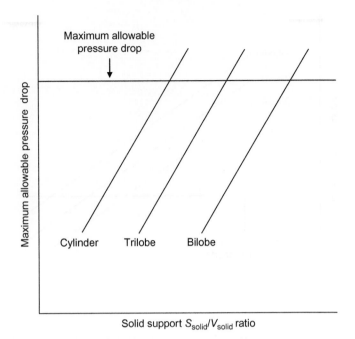

Figure 4.2 Maximum allowable pressure drop as a function of solid support S_{Solid}/V_{Solid}. (Adapted with permission from J. H. Worstell and J. H. Worstell, "Improve Fixed-bed Reactor Performance without Capital Expenditure", Chemical Engineering Progress, January 2004, pages 51–57.)

If the process underway in the fixed-bed reactor is pore diffusion rate limited, then

$$\frac{1}{k_{PD}(A_P/V_P)}$$

controls the rate of reactant consumption or product formation, depending which is monitored. In this case

$$k_{Overall} \propto k_{PD}(A_P/V_P)$$

Therefore, increasing A_P, which is the average cross-sectional area of the pores, increases diffusion along the pores; however, it has an upper limit. As average A_P increases, the average pore surface area decreases. Thus, the overall rate constant $k_{Overall}$ decreases due to a decrease in the number of active sites. Conversely, as average A_P decreases, the average pore surface area increases, which increases the number of active sites. However, resistance to diffusion along the pores increases, which causes a decline in $k_{Overall}$. Therefore, plotting $k_{Overall}$ as a function of A_P, or more generally, as average pore diameter D_P, produces a bell-shaped optimum, which must be determined experimentally.

Increasing A_P/V_P also increases the overall rate constant $k_{Overall}$. A_P/V_P is

$$\frac{(\pi/4)D_P^2}{(\pi/4)D_P^2 L_P} = \frac{1}{L_P}$$

where L_P is the pore length and D_P is pore diameter. L_P has two extensions: one across the diameter of the catalyst extrudate; the other along the axis of the catalyst extrudate. Therefore, shrinking the catalyst extrudate, i. e., increasing its S_{Solid}/V_{Solid}, reduces L_P, which increases the diffusion of reactant and product through the catalyst extrudate. The restrictions to maximizing A_P/V_P are the same as those for maximizing S_{Solid}/V_{Solid}.

Unfortunately, as important as bulk physical properties of the solid support are to improving the performance of a given fixed-bed reactor, they do not scale well from laboratory tests. Generally, we specify a minimum crush strength for a solid support. If a potential solid support meets that minimum crush strength, which has been proven to support its own weight in a commercial-sized fixed-bed reactor, then we agree to test the new solid support in a commercial fixed-bed reactor. Any laboratory measured crush strength below the commercial-scale "tested" minimum crush strength would not be charged to a fixed-bed reactor. The same comments apply to maximum values, such as attrition or powder generation.

If the process underway in the fixed-bed reactor is kinetics or reaction rate limited, then we have to change the chemistry of the solid-supported catalyst, which involves

1. identifying a different active site, generally a different metal, for catalyzing the desired reaction;
2. determining the proper quantity of metal per unit mass or per unit surface area of the solid support;
3. determining the optimal average pore diameter for the solid-supported catalyst;
4. determining the by-products produced by the solid-supported catalyst;
5. determining whether by-product formation occurs simultaneously or parallel to product formation;
6. optimizing product formation relative to by-product formation by adjusting the average pore diameter and surface area of the solid-supported catalyst;

7. determining the rate limiting step for product formation and for by-product formation;
8. determining the energy of activation for each known reaction, then optimizing process temperature so as to maximize product formation relative to by-product formation;
9. optimizing the activation procedure for the solid-supported catalyst;
10. determining whether the solid-supported catalyst deactivates via a spatially dependent or a temporally dependent mechanism;
11. optimizing the regeneration procedure for the solid-supported catalyst.

The above catalyst development program should not be initiated by the faint-hearted or by those with limited time or money. To complete a solid-supported catalyst development program requires considerable time and significant financial commitment.

4.3 IMPROVING PRODUCT SELECTIVITY—MONOFUNCTIONAL CATALYSTS

When managing the performance of a fixed-bed reactor, we are concerned with the amount of product produced by it relative to the amount of feed entering it and we are concerned about the amount of feed converted to an undesirable product relative to the desired product. The former concern is the "yield" of the fixed-bed reactor and the latter concern is the "selectivity" of the fixed-bed reactor.

Consider the reaction mechanism

$$A + AS \rightarrow P + AS$$
$$A + A + AS \rightarrow U + AS$$

where A is the reactant and AS is the active site on the solid-supported catalyst. These reactions occur concurrently, simultaneously on the solid-supported catalyst. P is the desired product and U is the undesired product or by-product. [A], [P], and [U] are the concentrations of each respective molecule and have units of $mols/m^3$. The top reaction has rate constant k_P while the bottom reaction has rate constant k_U; the unit for both rate constants is 1/s. Olefin isomerization on a solid acid is an example of such a mechanism: once the olefin interacts with the solid acid site, it can isomerize or it can react with a second A molecule to form an olefin dimer. The rate equations for this reaction mechanism are

$$R_P = k_P[A]^n$$
$$R_U = k_U[A]^m$$

The disappearance rate for A molecules is

$$-R_A = k_P[A]^n + k_U[A]^m$$

The yield of P is

$$\frac{R_P}{-R_A} = \frac{k_P[A]^n}{k_P[A]^n + k_U[A]^m}$$

and the yield of U is

$$\frac{R_P}{-RA} = \frac{k_U[A]^m}{k_P[A]^n + k_U[A]^m}$$

Therefore, the local selectivity of P to U is

$$\frac{R_P/-R_A}{R_U/-R_A} = \frac{k_P[A]^n/(k_P[A]^n + k_U[A]^m)}{k_U[A]^m/(k_P[A]^n + k_U[A]^m)}$$

Simplifying and rearranging yields

$$\frac{R_P}{R_U} = \frac{k_P[A]^n}{k_U[A]^m} = \left(\frac{k_P}{k_U}\right)[A]^{n-m}$$

If neither reaction is pore diffusion rate limited and $n = m$, then selectivity toward desired product is independent of solid support pore structure. If neither reaction is pore diffusion rate limited and $n < m$, then the last equation above suggests that adding an inert diluent to reduce the concentration of A improves selectivity toward desired product. However, the processing cost for producing P increases because inert diluent must subsequently be separated from desired product. If product formation is not pore diffusion rate limited but undesired or by-product formation is pore diffusion rate limited, then large diameter pellets or extrudates and small diameter pores increases selectivity toward desired product. If the first reaction is pore diffusion rate limited and the second reaction is not pore diffusion rate limited, then small diameter pellets or extrudates and large diameter pores increases selectivity toward desired product. If both reactions are pore diffusion rate limited and $k_P > k_U$ and $n = m$, then small diameter pellets or extrudates and large diameter pores improves selectivity toward desired product. If $k_P > k_U$ and $n < m$, then small diameter pores or adding an inert diluent to reduce [A] improves selectivity toward desired product. If $k_U > k_P$ and $n < m$, then the last equation above suggests adding an inert diluent to reduce the concentration of A improves selectivity toward desired product.[1-3]

Consider the reaction mechanism for parallel, competing reactions

$$A + AS \rightarrow P + AS$$
$$B + AS \rightarrow U + AS$$

Dehydrogenation of cyclohexane to cyclohexene in the presence of cyclopentane provides an example of this reaction mechanism. This example commonly occurs in petroleum reforming processes.[1] The selectivity for this reaction mechanism is

$$\frac{R_P}{R_U} = \frac{k_P[A]^n}{k_U[B]^m}$$

If neither reaction is pore diffusion rate limited and $n = m$, then selectivity toward desired product is independent of solid support pore structure. For this case, if $n > m$, then increasing [A] relative to [B] increases selectivity toward P and if $n < m$, then diluting B relative to A increases selectivity toward P. If the first reaction is not pore diffusion rate limited but the second reaction is pore diffusion rate limited, then increasing pellet or extrudate diameter and decreasing pore diameter increases selectivity toward P. If the first reaction is pore diffusion rate limited and the second reaction is not pore diffusion rate limited, then decreasing pellet or extrudate diameter and increasing pore diameter increases selectivity toward P. If both reactions are pore diffusion rate limited and $k_P > k_U$ and $n = m$, then decreasing pellet or extrudate diameter and increasing pore diameter increases selectivity toward desired product. The same is true for $k_P > k_U$ and $n > m$. If $k_P > k_U$ and $n < m$, then decreasing pellet or extrudate diameter and decreasing pore diameter increases selectivity toward P. And, if $k_U > k_P$ and $n < m$, then adding an inert diluent to reduce [B] improves selectivity toward desired product.

Consider the reaction mechanism describing consecutive reactions; it is

$$A + AS \rightarrow P + AS$$
$$P + AS \rightarrow U + AS$$

The rate expressions are

$$-R_A = k_P[A]^n$$
$$R_P = k_P[A]^n - k_U[P]^m$$
$$R_U = k_U[P]^m$$

and selectivity is

$$\frac{R_P}{R_U} = \frac{k_P[A]^n - k_U[P]^m}{k_U[P]^m} = \frac{k_P[A]^n}{k_U[P]^m} - 1$$

If neither reaction is pore diffusion rate limited and $n = m$, then selectivity toward desired product is independent of solid support pore structure. For this case, if $n > m$, then increasing [A] relative to [P] increases selectivity toward P and if $n < m$, then diluting P relative to A increases selectivity toward P. If the first reaction is pore diffusion rate limited and the second reaction is not pore diffusion rate limited, then decreasing pellet or extrudate diameter and increasing pore diameter increases selectivity toward P. If both reactions are pore diffusion rate limited and $k_P > k_U$ and $n = m$, then decreasing pellet or extrudate diameter and increasing pore diameter increases selectivity toward desired product. The same is true for $k_P > k_U$ and $n > m$. If $k_P > k_U$ and $n < m$, then decreasing pellet or extrudate diameter and decreasing pore diameter increases selectivity toward P. And, if $k_U > k_P$ and $n < m$, then adding an inert diluent to reduce [P] improves selectivity toward desired product.

In each of the above cases, thorough experimentation must be done to confirm the results achieved via ideal models and deductive logic.

4.4 IMPROVING PRODUCT SELECTIVITY—BIFUNCTIONAL CATALYSTS

Some chemical processes require two or more catalyzed reactions to produce the desired product. The two catalyzed reactions can be performed in series in separate fixed-bed reactors so long as the intermediate molecule is stable. On the other hand, fixed-bed reactor productivity and efficiency can be substantially improved if the two reactions are done in the same reactor, i.e., the fixed-bed reactor would contain both solid-supported catalysts. Operating such a fixed-bed reactor requires physically mixing the two solid-supported catalysts prior to loading them into the reactor. The size and shape of both solid-supported catalysts must be the same to ensure against subsequent catalyst segregation due to the loading procedure or due to fluid flow through the catalyst mass during operation. An alternative method for loading a fixed-bed reactor with different solid-supported catalysts is to layer the catalysts, i.e., for a liquid phase process, charge the fixed-bed reactor with the catalyst responsible for the first reaction, then load atop that catalyst, the second catalyst. The opposite is done for a gas phase process. This procedure essentially combines two separate fixed-bed reactors into one fixed-bed reactor. Both these methods limit the productivity and efficiency improvements attainable since each reaction occurs on separate solid-supported catalyst pellets or extrudates. We can obtain greater productivity and efficiency by placing both catalysts in multiple, alternating layers in one fixed-bed reactor. However, both these methods limit attainable productivity and efficiency improvements since each reaction occurs on separate solid-supported catalyst pellets or extrudates. Placing both catalytic active sites on the same solid-support achieves the maximum productivity and efficiency increase and solves the problem of potential particle segregation inherent in physically mixing two solid-supported catalysts.

In the late 1940s, Universal Oil Products (UOP) and others developed a bifunctional catalyst for petroleum reforming. That catalyst contained platinum for paraffin dehydrogenation and olefin hydrogenation on porous alumina, which provided the solid acid sites for converting linear olefins to iso-olefins.[4] However, platinum is also a good hydrogenolysis catalyst, thus its use as a hydrocracking catalyst.[5(p281)] During the 1960s, what is now ExxonMobil found that Group IB metals mixed with Group VIII metals greatly reduced the hydrogenolysis induced by bifunctional reforming catalysts.[6] This result led to the development of bifunctional, bimetallic reforming catalysts, which made reformed gasoline possible.

The hydrocarbon molecules in the gasoline boiling range are called "naphtha" and boil between 30°C and 200°C and encompass carbon numbers from five to twelve. In petroleum, these carbon numbers are mostly linear hydrocarbons. As such, they have low "octane numbers," i.e., they predetonate in front of the ignition wave occurring in an engine cylinder during combustion. This predetonation causes "engine knock," which damages engine bearings and other mechanical parts. Highly branched hydrocarbons and aromatic hydrocarbons have high octane numbers; thus, they do not cause engine knocking. From the 1920s until the mid-1960s, gasoline marketers added tetraethyl lead to gasoline to increase its antiknock characteristics. However, the volatilized lead compounds resulting from gasoline combustion is harmful to people and to the environment. Therefore, tetraethyl lead was banned from gasoline. In response, refining companies developed catalytic reforming, which takes linear naphtha hydrocarbon molecules and branches them to iso-hydrocarbon molecules and cyclizes and dehydrogenates a portion of them to aromatics—all of which possess high octane numbers. Thus, catalytic reforming changes the molecular structure of naphtha hydrocarbons but not their boiling point range.

Bifunctional solid-supported catalysts are used in other processes as well. For example, bifunctional, solid-supported catalysts are used for the partial oxidation of hydrocarbons, such as propylene to acrolein; for the ammoxidation of propylene to acrylonitrile; for hydrodesulfurization of petroleum; and, for the hydrodenitrogenation of petroleum.

In general, a solid support provides strong acid sites for reaction with hydrocarbon molecules and surface area for attachment of and dispersion of a transition metal or a mixture of metals. If neither reaction is pore diffusion rate limited, then increasing pellet or extrudate diameter and decreasing pore diameter improves product formation rate. If the initial dehydrogenation reaction and subsequent hydrogenation reaction are pore diffusion rate limited and the olefin isomerization reaction at the acid site is not pore diffusion rate limited, then decreasing pellet or extrudate diameter and increasing pore diameter improves product formation rate. For this case, placing most of the dehydrogenation/hydrogenation sites in the outer third of the pellet or extrudate also improves product formation rate. If the dehydrogenation/hydrogenation reactions are not pore diffusion rate limited and the olefin isomerization reaction is pore diffusion rate limited, then increasing pellet or extrudate diameter and increasing pore diameter improves product formation rate. Placing most of the dehydrogenation/hydrogenation sites in the outer third of the pellet or extrudate also improves product formation rate. If all reactions are pore diffusion rate limited, then decreasing pellet or extrudate diameter, increasing pore diameter, and placing most of the dehydrogenation/hydrogenation sites in the outer third of the pellet or extrudate increases product formation rate.

In each of the above cases, thorough experimentation must be done to confirm the results achieved via ideal models and deductive logic.

4.5 SUMMARY

This chapter discussed methods for improving the productivity and efficiency of a solid-supported catalyst. It began by stressing the need to identify the rate limiting step of the chemical process before planning and initiating a solid-supported catalyst improvement program. This chapter also presented methods for identifying the rate limiting step of a solid-supported catalytic process, then discussed how to improve reactant consumption or product formation through catalyst modification. It concluded with methods for improving the selectivity of solid-supported catalysts.

REFERENCES

1. Thomas J, Thomas W. *Principles and practice of heterogeneous catalysis.* Weinheim, Germany: VCH; 1997. p. 419.

2. Rase H. *Chemical reactor design for process reactors—volume one: principles and techniques.* New York, NY: John Wiley & Sons; 1977. p. 279.

3. Worstell J. Don't Act Like a Novice About Reaction Engineering. *Chem Eng Prog* 2001; 68–72.

4. Haensel V. Reforming Approaches the Ideal Solution for Low-octane and Straightrun Gasolines. *Oil Gas J* 1950;**30**(March):82–7.

5. Gates B, Katzer J, Schuit G. *Chemistry of catalytic processes.* New York, NY: McGraw-Hill Book Company; 1979. p. 279.

6. Sinfelt J. *Bimetallic catalysts: discoveries, concepts, and applications.* New York, NY: John Wiley & Sons; 1983 [chapter 2].

CHAPTER 5

Scaling Fixed-Bed Reactors

5.1 INTRODUCTION

Fixed-bed reactors come in all sizes, but we generally group them as laboratory-scale, pilot plant-scale, or commercial-scale. We operate laboratory-scale fixed-bed reactors when developing a new process, investigating a new solid-supported catalyst, qualifying for commercial use a different catalyst, and supporting an existing commercial process.

We use pilot plant-scale fixed-bed reactors when developing a new process or supporting an existing commercial process. Depending on the process, we may qualify a new or different solid-supported catalyst in a pilot plant; this usually occurs when we want to qualify the test catalyst using commercial plant feeds. We also use pilot plant-scale fixed-bed reactors to determine the contractual performance criteria of a given solid-supported catalyst. This last use of pilot plant-scale fixed-bed reactors is done for catalysts containing precious metals or for catalysts licensed with a royalty fee. Silver-containing solid-supported, ethylene oxide catalysts are an example of catalysts marketed per pilot plant performance criteria.

We operate commercial-scale fixed-bed reactors to produce product for market.

Upscaling involves moving a catalyst or process from the laboratory, through a pilot plant, to a commercial plant. We upscale when developing a new process or a new catalyst. Downscaling occurs for existing commercial processes that are old enough to have had their original pilot plants dismantled. But, a time comes for such commercial processes when it is desirable to build a new pilot plant for solving current operating problems.

Upscaling and downscaling require models. We use these models to reduce the time spent experimenting at the laboratory-scale and the time spent validating at the pilot plant-scale, which ultimately reduces the cost of the research program. The major cost savings from modeling come from not building a nonfunctional commercial-scale fixed-bed reactor or an inappropriate pilot plant-scale fixed-bed reactor. A nonfunctional commercial-scale fixed-bed reactor is one that does not produce product meeting published specifications or one that does not produce product at an economic rate. An inappropriate pilot plant-scale fixed-bed reactor is one that operates, unbeknownst to us, in a process regime different from that of the commercial-scale fixed-bed reactor. Thus, the importance of models.

5.2 MODELS

There are four types of models. They are:

1. true models;
2. adequate models;
3. distorted models;
4. dissimilar models.

True models involve building all significant process features to scale. Thus, the model is an exact replica of the commercial plant, which we call the "prototype." We build true models in some safety investigations when determining the cause of a grievous event. While true models may provide highly accurate information, they are capital intensive, expensive to operate, and require extended time to build.

Adequate models predict one characteristic of the prototype accurately. If the sizes of the model and the prototype are significantly different, then it is unlikely that we can achieve complete similarity. And, for complex processes, a complete model is actually a full-scale prototype, i.e., a true model.[1] When the modeled characteristic is the dominant, controlling factor in the process, then an adequate model may be sufficient. For example, processes using a solid-supported catalyst are generally stagnant film diffusion rate limited or pore diffusion rate limited. If a process is so limited, then we only have to ensure the same controlling regime in our laboratory or pilot plant reactors. If we do not ensure equivalent controlling regimes in the laboratory or pilot plant reactors, then any process development or process support will be wasted effort If we do not consider whether the commercial process is stagnant film diffusion limited, pore diffusion rate limited, or reaction rate limited, then we will finish our effort with an expensive scattergram of the experimental results. This situation actually occurs more often than we like to admit. Many commercial processes using a solid-supported catalyst are pore diffusion rate limited due to high interstitial fluid velocity through the catalyst mass. Such high fluid velocity minimizes the boundary layer surrounding each catalyst pellet, thereby making the process pore diffusion rate limited. Unfortunately, most pilot plant processes, i.e., models, using porous solid catalysts are operated at low interstitial fluid velocities in order to minimize feed and product volumes at the research site. Both can be a safety hazard if they are hydrocarbons and the product can be a disposal issue since it cannot be sold and is generally not fed into a commercial process. In such situations, stagnant film diffusion rate is the controlling regime. The result of a multiyear, multicatalyst testing effort will be an expensive scattergram around the average value for the stagnant film diffusion rate constant. On the other hand, considerable effort can be made at the laboratory scale to ensure that catalyst testing occurs in the reaction rate limited regime. Plots with impressive correlations result from these types of experimental programs. Unfortunately, when the best catalyst is tested in the prototype, it displays the same efficiency and productivity as the current catalyst. In such cases, the prototype is either stagnant film diffusion rate or pore diffusion rate limited. It does not matter how reactive the catalyst is in the laboratory; in the prototype, the process is incapable of keeping the catalytic active site saturated with reactant. In conclusion, the controlling regime of the model must be identical to the controlling regime of the prototype. With regard to the process, adequate models behave similarly to their prototypes, even though they may be many times smaller than their prototypes.

In distorted models, we violate design conditions intentionally for one reason or another. Such distortion affects the prediction equation. In other words, we have to correct data from the model in order to simulate the prototype. Hydrologic river basin models are the most common distorted models. In these models, the horizontal and vertical lengths do not have the same ratios or "scaling factors." In a geometrically similar model, the horizontal and vertical ratios are equal, for example

$$\frac{{}_P L_H}{{}_M L_H} = \kappa \quad \text{and} \quad \frac{{}_P L_V}{{}_M L_V} = \kappa$$

where ${}_P L_H$ is the prototype horizontal length of interest; ${}_M L_H$ is the model horizontal length equivalent to ${}_P L_H$; ${}_P L_V$ is the prototype vertical length of interest; and $_M L_V$ is the model vertical length equivalent to $_P L_V$. κ and λ are constants; they are "scaling factors." For a distorted model

$$\frac{_P L_H}{_M L_H} = \kappa \quad \text{and} \quad \frac{_P L_V}{_M L_V} = \lambda$$

where $\kappa \neq \lambda$. It is "legal" to use distorted models, so long as we know we are doing it and we understand why we are doing it. With regard to the process, distorted models behave in a manner similar to their prototypes; however, one dimension of the model will not be scaled equivalently to the other dimensions. Thus, a distorted model may look squat or tall or broad, depending on the distortion, when compared to its prototype.

Dissimilar models comprise the fourth and last model type. Such models have no apparent resemblance to the prototype. Dissimilar models have, as their name states, no similarity to their prototypes. These models provide information about the prototype through suitable analogies.

5.3 SIMILARITY

We base our models on similarity. Four similarities are important to chemical engineers. They are:

1. geometrical;
2. mechanical;
3. thermal;
4. chemical.

In general, geometric similarity means that given two objects of different size, if there is a point within the smaller object, which we identify as the model, with coordinates x_M, y_M, and z_M, and a similar point within the larger object, i.e., the prototype, with coordinates x_P, y_P, and z_P, then the two objects are similar at that given point if

$$\frac{x_P}{x_M} = \frac{y_P}{y_M} = \frac{z_P}{z_M} = L$$

The two objects are geometrically similar if the above condition holds for all corresponding points within the two objects.

Mechanical similarity comprises three subsimilarities, which are static similarity, kinematic similarity, and dynamic similarity. Static similarity demands that two geometrically similar objects have relative deformation for a constant applied stress. This similarity is of interest to civil and structural engineers.

Kinematic similarity means the constituent parts of a model and prototype mechanism or process in translation follow similar paths or streamlines if the model and prototype are geometrically similar. Thus

$$\frac{v_P}{v_M} = V$$

where v_M is the velocity of the translating model part or particle and v_P is the translating prototype part or particle. V is the velocity scaling factor.

Dynamic similarity demands the ratio of the forces inducing acceleration be equal at corresponding locations in geometrically similar mechanisms or processes. In other words, the ratio

$$\frac{F_P}{F_M} = F$$

where F_M is the force at location x_M, y_M, and z_M in the model and F_P is the force at location x_P, y_P, and z_P in the prototype, holds true at every corresponding location in the two mechanisms or processes.

Thermal similarity occurs when the ratio of the temperature difference at corresponding locations of a geometrically similar mechanism or process are equal. If translation, i.e., movement, occurs, then the process must also demonstrate kinematic similarity for thermal similarity to occur. Thus, thermal similarity requires geometric similarity and kinematic similarity.

As chemical engineers, our major concern is the reactions occurring in the process. We want our model to reflect what occurs in our prototype. To ensure that outcome, our model must be chemically similar to our prototype. Chemical similarity demands the ratio of concentration differences at all corresponding locations in the model and in the prototype be equal. Therefore, our model and prototype must also be geometrically, mechanically, and thermally similar.

Consider two mechanical processes involving the Navier–Stokes equation. Let one process be large and the other process be small. Our question: is the larger process similar to the smaller process? The best way to answer our question is to convert the Navier–Stokes equation into a dimensionless form. To do that, define a characteristic length L and velocity V, then form the dimensionless variables x^* and v^*, which are

$$x_S^* = \frac{x_S}{L} \quad \text{and} \quad v_{x,S}^* = \frac{v_{x,S}}{V}$$

where the subscript S identifies the small process; x, the length in the x-direction; L, the characteristic length; x_S^*, the dimensionless length in the x-direction; $v_{x,S}$ the fluid velocity in the x-direction in the small process; V, the characteristic velocity; and, $v_{x,S}^*$, the dimensionless velocity in the x-direction. We define dimensionless pressure as

$$p^* = \frac{p}{\rho V^2}$$

and we define dimensionless time as

$$t^* = \frac{Vt}{L}$$

The Navier–Stokes equation in one-dimension for the small process is

$$\frac{\partial v_{x,S}}{\partial t} + v_{x,S}\frac{\partial v_{x,S}}{\partial x_S} = -\frac{1}{\rho}\frac{\partial p_S}{\partial x_S} + g_x + \frac{\mu}{\rho}\frac{\partial^2 v_{x,S}}{\partial (x_S)^2}$$

where p_S is the pressure of the small process; g_x is acceleration due to gravity; μ is fluid dynamic viscosity; and ρ is fluid density; $\partial v_{x,S}/\partial t$ represents the local acceleration of the fluid particle; $v_{x,S}(\partial v_{x,S}/\partial x_S)$ is the convective acceleration of the fluid particle; $(1/\rho)(\partial p_S)/(\partial x_S)$ represents the pressure acceleration due to pumping action; and $(\mu/\rho)(\partial^2 v_{x,S}/\partial (x_S)^2)$ is the viscous deceleration generated by objects in the fluid's flow path.[2] Converting the dimensional equation to a dimensionless equation yields

$$\left(\frac{V^2}{L}\right)\frac{\partial v_{x,S}^*}{\partial t^*} + \left(\frac{V^2}{L}\right)v_{x,S}^*\frac{\partial v_{x,S}^*}{\partial x_S^*} = -\left(\frac{V^2}{L}\right)\frac{\partial p_S^*}{\partial x_S^*} + g_x + \frac{\mu V}{\rho L^2}\frac{\partial^2 v_{x,S}^*}{\partial (x_S^*)^2}$$

Multiplying the above equation by L/V^2 and simplifying yields

$$\frac{\partial v_{x,S}^*}{\partial t^*} + v_{x,S}^*\frac{\partial v_{x,S}^*}{\partial x_S^*} = -\frac{\partial p_S^*}{\partial x_S^*} + \frac{g_x L}{V^2} + \frac{\mu}{\rho L V}\left(\frac{\partial^2 v_{x,S}^*}{\partial (x_S^*)^2}\right)$$

Now consider the one-dimensional Navier–Stokes equation for the larger process, identified by the subscript L; it is

$$\frac{\partial v_{x,L}}{\partial t} + v_{x,L}\frac{\partial v_{x,L}}{\partial x_L} = -\frac{1}{\rho}\frac{\partial p_L}{\partial x_L} + g_x + \frac{\mu}{\rho}\frac{\partial^2 v_{x,L}}{\partial (x_L)^2}$$

We can convert this Navier–Stokes equation into a dimensionless equation just as before. Doing so gives us

$$\frac{\partial v_{x,L}^*}{\partial t^*} + v_{x,L}^*\frac{\partial v_{x,L}^*}{\partial x_L^*} = -\frac{\partial p_L^*}{\partial x_L^*} + \frac{g_x L}{V^2} + \frac{\mu}{\rho L V}\left(\frac{\partial^2 v_{x,L}^*}{\partial (x_L^*)^2}\right)$$

Note both dimensionless equations have the same dimensionless groups, namely

$$\frac{g_x L}{V^2} \quad \text{and} \quad \frac{\mu}{\rho L V}$$

which are the inverse Froude number and the inverse Reynolds number. The Froude number is the ratio of the inertial forces to gravitational forces and the Reynolds number is the ratio of the inertial forces to viscous forces. Thus, if

$$\left(\frac{g_x L}{V^2}\right)_S = \left(\frac{g_x L}{V^2}\right)_L$$

and

$$\left(\frac{\mu}{\rho L V}\right)_S = \left(\frac{\mu}{\rho L V}\right)_L$$

then the two processes are mechanically equivalent.

However, the two processes must be geometrically similar for them to be mechanically similar. For each process, we defined

$$x_S^* = \frac{x_S}{L} \quad \text{and} \quad x_L^* = \frac{x_L}{L}$$

Thus

$$L = \frac{x_S}{x_S^*} \quad \text{and} \quad L = \frac{x_L}{x_L^*}$$

Equating the above equations, then rearranging give us

$$\frac{x_S}{x_S^*} = \frac{x_L}{x_L^*}$$

$$\frac{x_S}{x_L} = \frac{x_S^*}{x_L^*}$$

Therefore, the two processes are geometrically similar.

In summary, two processes are similar if their dimensionless geometric ratios are equal and if their dimensionless process parameters are equal. In other words, each process will generate a set of dimensionless parameters denoted by Π. When corresponding parameters are equal, then the comparator processes are similar. Symbolically

$$\Pi_1^{Geometric} = \Pi_2^{Geometric}$$

$$\Pi_1^{Static} = \Pi_2^{Statiic}$$

$$\Pi_1^{Kinematic} = \Pi_2^{Kinematic}$$

$$\Pi_1^{Dynamic} = \Pi_2^{Dynamic}$$

$$\Pi_1^{Thermal} = \Pi_2^{Thermal}$$

$$\Pi_1^{Chemical} = \Pi_2^{Chemical}$$

Thus, similarity rests upon dimensional analysis.

5.4 THEORY OF MODELS

The most general equation for a prototype is

$$\Pi_1^P = f(\Pi_2^P, \Pi_3^P, \ldots, \Pi_n^P)$$

where the subscript numeral identifies a dimensionless parameter and superscript P indicates prototype. This equation applies to all mechanisms or processes that are comprised of the same dimensional variables. Thus, it applies to any model of the same mechanism or process, which means we can write a similar equation for that model

$$\Pi_1^M = f(\Pi_2^M, \Pi_3^M, \ldots, \Pi_n^M)$$

Dividing the prototype equation by the model equation gives us

$$\frac{\Pi_1^P}{\Pi_1^M} = \frac{f(\Pi_2^P, \Pi_3^P, \ldots, \Pi_n^P)}{f(\Pi_2^M, \Pi_3^M, \ldots, \Pi_n^M)}$$

Note that if $\Pi_2^P = \Pi_2^M$ and $\Pi_3^P = \Pi_3^M$ and so on, then

$$\frac{\Pi_1^P}{\Pi_1^M} = 1$$

Thus, $\Pi_1^P = \Pi_1^M$, which is the condition for predicting prototype behavior from model behavior. The conditions

$$\Pi_2^P = \Pi_2^M$$
$$\Pi_3^P = \Pi_3^M$$
$$\vdots \quad \vdots$$
$$\Pi_n^P = \Pi_n^M$$

constitute the design specifications for the prototype from the model or the model from the prototype, depending whether we are upscaling or downscaling. If all these conditions are met, then we have a true model. If the above conditions hold for the controlling regime of the model and the prototype, then we have an adequate model. If most of the above conditions hold, then we have a distorted model that requires a correlation to relate Π_1^P and Π_1^M; in other words, we need an additional function such that

$$\Pi_1^P = f(\text{correlation})\Pi_1^M$$

If none of the above conditions holds true, then we have an analogous model.

We generally do not build true models in the CPI because the processes are so complex. A true model of a chemical process implies building a commercial-sized plant, which is far too costly and time consuming for an organization to do. Most models in the CPI are adequate or distorted models. Of these two types, adequate models are the better since they model the controlling regime of the process. Distorted models are more difficult to use because we have to determine the correlation between the distorted model and the prototype. Developing that correlation takes time and costs money ... two commodities in short supply in our global economy.

5.5 SCALING ADIABATIC FIXED-BED REACTORS

To successfully downscale or upscale a chemical process, we must identify the pertinent dimensionless parameters for that chemical process, then ensure that the dimensionless parameters for the downscaled or upscaled process equal those of the reference process. The two chemical processes will operate similarly when their dimensionless parameters are equal. The same is true when downscaling or upscaling a fixed-bed reactor.

We must use dimensional analysis to determine the pertinent dimensionless parameters when downscaling or upscaling a fixed-bed reactor. Those engineering disciplines concerned with fluid flow, such as aeronautical, civil, and mechanical, have used dimensional analysis to good effect. Their success is largely attributable to the fact that fluid flow requires only three fundamental dimensions and generates a limited number of dimensionless parameters. Thus, the algebra is amenable to hand calculation.

Mechanical and chemical engineers are both concerned about heat flow, either into or from a given mechanism or process. Working with heat flow requires a fourth fundamental dimension, namely, temperature or thermal energy, which complicates the algebra of dimensional analysis. The situation is further complicated by the flow of fluid initiated by or required by heat transfer, thereby requiring more dimensionless parameters to fully describe the mechanism or process. And, where complicated algebra occurs, mistakes happen. Dimensional analysis involving four fundamental dimensions has been done many times by hand, but such efforts involve significant amounts of time and effort to obtain the first solution, then to check that solution for possible algebraic errors. Thus, the application of dimensional analysis to situations involving heat transfer is much smaller than for those situations involving fluid flow.

The situation is even more complicated for chemical engineers, who are concerned with chemical change and with producing chemical products at acceptable rates. Analyzing chemical processes requires a fifth fundamental dimension, that dimension being "amount of substance," which is moles in the SI system of units. Chemical change also involves fluid flow and heat transfer, either initiated by the chemical reaction itself or required by the chemical process. Thus, the algebra for dimensional analysis of chemical processes is daunting. Due to the algebraic complexity of dimensional analysis, chemical engineers have not utilized it to the extent that other engineering disciplines have employed it.

The matrix formulation of dimensional analysis and the availability of free-for-use matrix calculators on the Internet resolve the algebraic issues for chemical engineers and provide a rapid method for determining the dimensionless parameters best describing a chemical process.[3] See the Appendix to this book for the matrix format of dimensional analysis.

It is difficult to maintain strict similarity in chemical processes due to the number of variables involved. This point is especially true for fixed-bed reactors. For example, we do not change the size or shape of the solid-supported catalyst when scaling a fixed-bed reactor. Thus, the ratio of d_P/D, where d_P is the diameter of the catalyst pellet or extrudate and D is the reactor diameter, varies from the reference reactor to the scaled reactor. In this case, geometric similarity does not hold when we scale fixed-bed reactors. However, knowledge of this fact should not hinder our use of dimensional analysis when scaling fixed-bed reactors. We simply need to remember it at all times during the scaling effort and make an estimate of its impact on the final effort. d_P/D impacts the distribution of fluid flowing through the fixed-bed reactor. The void fraction of the solid-supported catalyst at the wall is one; therefore, fluid velocity along the wall is high relative to the fluid velocity through the catalyst mass. For large diameter fixed-bed reactors, the fluid velocity profile across the catalyst mass is essentially flat: fluid flow along the reactor wall has little impact on the overall performance of the reactor. However, as D decreases, i.e., as d_P/D increases, fluid flow along the reactor wall begins to impact the fluid velocity profile across the catalyst mass, which can impact the overall performance of the fixed-bed reactor.[4] A rule of thumb exists, which states fixed-bed reactor performance is independent of d_P/D if $D > 10d_P$. In other words, fluid flow along the wall of the fixed-bed reactor can be neglected during a scaling design if the reactor diameter is greater than 10 solid-supported catalyst diameters. Such a design will not be strictly similar, geometrically, but it will produce a fixed-bed reactor that performs similarly to its reference reactor.

When scaling a fixed-bed reactor, it may be difficult to maintain chemical similarity. $k_{overall}$ represents

- the movement of reactant molecules through the stagnant film surrounding each solid-supported catalyst pellet or extrudate in the fixed-bed reactor;

- the movement of reactant molecules along a single catalyst pore;
- the conversion of reactant molecules to product molecules at the active site.

If expressed as product formation, $k_{overall}$ represents the above progression in reverse. Note that each of the above steps can be rate-limiting. And, which step is rate-limiting shifts with scale. In general, laboratory-scale fixed-bed reactors are stagnant film diffusion rate limited; pilot plant-scale fixed-bed reactors are stagnant film or pore diffusion rate limited; and, commercial-scale fixed-bed reactors are pore diffusion or reaction rate limited. Thus, chemical similarity will most likely not hold when scaling a fixed-bed reactor. Again, this fact should not hinder our use of dimensional analysis when scaling a fixed-bed reactor, so long as we remember that the rate controlling step represented by $k_{overall}$ shifts with scale.

Consider a cylindrical tower filled with a solid-supported catalyst. The feed is liquid and flows upward through the reactor. The reactor operates adiabatically. The geometric variables are reactor diameter D [L]; reactor length L [L], which is the height of the catalyst mass; and solid-supported catalyst pellet or extrudate diameter d_P [L]. The material variables are fluid viscosity μ [$L^{-1}MT^{-1}$], fluid density ρ [$L^{-3}M$], fluid–solid heat capacity C_P [$L^2MT^{-2}\theta^{-1}$], fluid–solid heat conductivity k [$LMT^{-3}\theta^{-1}$], and molecular diffusivity D_{Diff} [L^2T^{-1}]. The process variables are:

- reactant concentration entering the reactor C_{In} and reactant concentration exiting the reactor C_{Out} [$L^{-3}N$];
- the heat of reaction $C_{In} \Delta H_R$ [$L^{-1}MT^{-2}$], where ΔH_R has dimensions of $L^2MT^{-2}N^{-1}$ and C_{In} has dimensions of [$L^{-3}N$];
- the interstitial fluid velocity through the reactor v [LT^{-1}]—interstitial velocity is $v = Q/\varepsilon A$, where Q is volumetric flow rate, A is the cross-sectional area of the empty cylindrical tower, and ε is the void fraction of the porous solid catalyst;
- the fluid temperature entering the reactor K_{In} [θ]—we determine all physical properties at K_{In};
- the temperature difference between the entering fluid and exiting fluid ΔK_{IO} [θ];
- the overall rate constant k_O [T^{-1}].

The dimensional table is

Variable		L	D	d_P	v	D_{Diff}	k_O	K_{In}	ΔK_{IO}	$C_{In}\Delta H_R$	C_{Out}	C_{In}	ρ	μ	C_P	k
Dimension	L	1	1	1	1	2	0	0	0	−1	−3	−3	−3	−1	2	1
	M	0	0	0	0	0	0	0	0	1	0	0	1	1	1	1
	T	0	0	0	−1	−1	−1	0	0	−2	0	0	0	−1	−2	−3
		0	0	0	0	0	0	1	1	0	0	0	0	0	−1	−1
	N	0	0	0	0	0	0	0	0	0	1	1	0	0	0	0

and the dimension matrix is

$$\begin{bmatrix} 1 & 1 & 1 & 1 & 2 & 0 & 0 & 0 & -1 & -3 & -3 & -3 & -1 & 2 & 1 \\ 0 & 0 & 0 & 0 & 0 & 0 & 0 & 0 & 1 & 0 & 0 & 1 & 1 & 1 & 1 \\ 0 & 0 & 0 & -1 & -1 & -1 & 0 & 0 & -2 & 0 & 0 & 0 & -1 & -2 & -3 \\ 0 & 0 & 0 & 0 & 0 & 0 & 1 & 1 & 0 & 0 & 0 & 0 & 0 & -1 & -1 \\ 0 & 0 & 0 & 0 & 0 & 0 & 0 & 0 & 0 & 1 & 1 & 0 & 0 & 0 & 0 \end{bmatrix}$$

The largest square matrix for this dimension matrix is 5×5; it is

$$R = \begin{bmatrix} -3 & -3 & -1 & 2 & 1 \\ 0 & 1 & 1 & 1 & 1 \\ 0 & 0 & -1 & -2 & -3 \\ 0 & 0 & 0 & -1 & -1 \\ 1 & 0 & 0 & 0 & 0 \end{bmatrix}$$

Its determinant is

$$|R| = \begin{vmatrix} -3 & -3 & -1 & 2 & 1 \\ 0 & 1 & 1 & 1 & 1 \\ 0 & 0 & -1 & -2 & -3 \\ 0 & 0 & 0 & -1 & -1 \\ 1 & 0 & 0 & 0 & 0 \end{vmatrix} = 3$$

Since $|R|$ is 3, the rank of this dimension matrix is 5 because it is a 5×5 square matrix. Therefore, the number of dimensionless parameters is

$$N_P = N_{Var} - R = 15 - 5 = 10$$

The above equation is the Buckingham's Pi Theorem which comes from the application of matrix algebra to process analysis.

The inverse of R is

$$R^{-1} = \begin{bmatrix} -3 & -3 & -1 & 2 & 1 \\ 0 & 1 & 1 & 1 & 1 \\ 0 & 0 & -1 & -2 & -3 \\ 0 & 0 & 0 & -1 & -1 \\ 1 & 0 & 0 & 0 & 0 \end{bmatrix}^{-1} = \begin{bmatrix} 0 & 0 & 0 & 0 & 1 \\ -0.33 & 0 & 0.33 & -1.33 & -1 \\ 0.33 & 1 & -0.33 & 2.33 & 1 \\ 0.33 & 1 & 0.66 & -0.66 & 1 \\ -0.33 & -1 & -0.66 & -0.33 & -1 \end{bmatrix}$$

and the bulk matrix is

$$B = \begin{bmatrix} 1 & 1 & 1 & 1 & 2 & 0 & 0 & 0 & -1 & -3 \\ 0 & 0 & 0 & 0 & 0 & 0 & 0 & 0 & 1 & 0 \\ 0 & 0 & 0 & -1 & -1 & -1 & 0 & 0 & -2 & 0 \\ 0 & 0 & 0 & 0 & 0 & 0 & 1 & 1 & 0 & 0 \\ 0 & 0 & 0 & 0 & 0 & 0 & 0 & 0 & 0 & 1 \end{bmatrix}$$

Therefore, $-R^{-1}*B$ is

$$-R^{-1}*B = \begin{bmatrix} 0 & 0 & 0 & 0 & 0 & 0 & 0 & 0 & 0 & -1 \\ 0.33 & 0.33 & 0.33 & 0.66 & 1 & 0.33 & 1.33 & 1.33 & 0.33 & 0 \\ -0.33 & -0.33 & -0.33 & -0.66 & -1 & -0.33 & -2.33 & -2.33 & -1.33 & 0 \\ -0.33 & -0.33 & -0.33 & 0.33 & 0 & 0.66 & 0.66 & 0.66 & 0.66 & 0 \\ 0.33 & 0.33 & 0.33 & -0.33 & 0 & -0.66 & 0.33 & 0.33 & 0.66 & 0 \end{bmatrix}$$

We can now assemble the total matrix, which is shown below. Each of the first 10 columns from the left bracket comprises a dimensionless parameter and is identified as such along the top of the total matrix. The dimensionless parameters, reading down the Π_i columns of the total matrix, are

$$T = \begin{bmatrix} L \\ D \\ d_P \\ v \\ D_{\text{Diff}} \\ k_O \\ K_{\text{In}} \\ K_{IO} \\ C_{\text{In}}\Delta H_R \\ C_{\text{out}} \\ C_{\text{In}} \\ \rho \\ \mu \\ C_P \\ k \end{bmatrix}$$

	Π_1	Π_2	Π_3	Π_4	Π_5	Π_6	Π_7	Π_8	Π_9	Π_{10}					
L	1	0	0	0	0	0	0	0	0	0	0	0	0	0	0
D	0	1	0	0	0	0	0	0	0	0	0	0	0	0	0
d_P	0	0	1	0	0	0	0	0	0	0	0	0	0	0	0
v	0	0	0	1	0	0	0	0	0	0	0	0	0	0	0
D_{Diff}	0	0	0	0	1	0	0	0	0	0	0	0	0	0	0
k_O	0	0	0	0	0	1	0	0	0	0	0	0	0	0	0
K_{In}	0	0	0	0	0	0	1	0	0	0	0	0	0	0	0
K_{IO}	0	0	0	0	0	0	0	1	0	0	0	0	0	0	0
$C_{\text{In}}\Delta H_R$	0	0	0	0	0	0	0	0	1	0	0	0	0	0	0
C_{out}	0	0	0	0	0	0	0	0	0	1	0	0	0	0	0
C_{In}	0	0	0	0	0	0	0	0	0	-1	0	0	0	0	-1
ρ	0.33	0.33	0.33	0.66	1	0.33	1.33	1.33	0.33	0	0.33	0	-0.33	1.33	1
μ	-0.33	-0.33	-0.33	-0.66	-1	-0.33	-2.33	-2.33	-1.33	0	-0.33	-1	0.33	-2.33	-1
C_P	-0.33	-0.33	-0.33	0.33	0	0.66	0.66	0.66	0.66	0	-0.33	-1	-0.66	0.66	-1
k	0.33	0.33	0.33	-0.33	0	-0.66	0.33	0.33	-0.66	0	0.33	1	0.66	0.33	1

$$\Pi_1 = \frac{L\rho^{0.33}k^{0.33}}{\mu^{0.33}C_P^{0.33}} \qquad \Pi_2 = \frac{D\rho^{0.33}k^{0.33}}{\mu^{0.33}C_P^{0.33}} \qquad \Pi_3 = \frac{d_P\rho^{0.33}k^{0.33}}{\mu^{0.33}C_P^{0.33}}$$

$$\Pi_4 = \frac{v\rho^{0.66}C_P^{0.33}}{\mu^{0.66}k^{0.33}} \qquad \Pi_5 = \frac{D_{\text{Diff}}\rho}{\mu} \qquad \Pi_6 = \frac{k_O\rho^{0.33}C_P^{0.66}}{\mu^{0.33}k^{0.66}}$$

$$\Pi_7 = \frac{K_{\text{In}}\rho^{1.33}C_P^{0.66}}{\mu^{2.33}} \qquad \Pi_8 = \frac{\Delta K_{IO}\rho^{1.33}C_P^{0.66}}{\mu^{2.33}} \qquad \Pi_9 = \frac{(C_S\Delta H_R)\rho^{0.33}C_P^{0.66}}{\mu^{1.33}k^{0.66}}$$

$$\Pi_{10} = \frac{C_{\text{Out}}}{C_{\text{In}}}$$

Since the above dimensionless parameters are independent of each other, we can multiply and divide them to remove the fractional indices. Combining dimensionless parameters in order to remove the fractional indices gives

$$\frac{\Pi_1}{\Pi_2} = \frac{L}{D} \qquad \frac{\Pi_3}{\Pi_2} = \frac{d_P}{D} \qquad \Pi_2\Pi_4 = \frac{\rho Dv}{\mu} = Re$$

$$\Pi_5 = \frac{\rho D_{\text{Diff}}}{\mu} = Sc^{-1} \qquad \frac{\Pi_1\Pi_6}{\Pi_4} = \frac{Lk_O}{v} = \tau k_O \qquad \frac{\Pi_1^2\Pi_6}{\Pi_5} = \frac{L^2 k_O}{D_{\text{Diff}}}$$

$$\frac{\Pi_1\Pi_6\Pi_9}{\Pi_4\Pi_7} = \frac{(C_{\text{In}}\Delta H_R)Lk_O}{\rho k v K_{\text{In}}} \quad \frac{\Pi_1^2\Pi_6\Pi_9}{\Pi_7} = \frac{(C_{\text{In}}\Delta H_R)L^2 k_O}{k K_{\text{In}}} \quad \frac{\Pi_8}{\Pi_7} = \frac{\Delta K_{IO}}{K_{\text{In}}}$$

$$\Pi_{10} = \frac{C_{\text{Out}}}{C_{\text{In}}}$$

Π_1/Π_2 is the "aspect ratio" of the reactor. Π_3/Π_2 is the ratio of solid-supported catalyst diameter to reactor diameter. $\Pi_2\Pi_4$ is the Reynolds number; Π_5 is the inverse Schmidt number, which is the ratio of momentum diffusivity and molecular diffusivity. $\Pi_1\Pi_6/\Pi_4$ is average residence time that a reactant molecule spends in the reactor; it is also the Group I Damkohler number Da^I, which is the ratio of chemical reaction rate to bulk flow rate. $\Pi_1^2\Pi_6/\Pi_5$ is the Group II Damkohler number Da^{II}, which is the ratio of chemical reaction rate to molecular to diffusion rate. $\Pi_1^2\Pi_6\Pi_9/\Pi_7$ is the Group IV Damkohler number Da^{IV}, which is the ratio of heat liberated or consumed to conductive heat transfer. We must multiply the dimensionless parameter $\Pi_1\Pi_6\Pi_9/\Pi_4\Pi_7$ by C_P/C_P, i.e., by one, to obtain

$$\frac{(C_{\text{In}}\,\Delta H_R)k_O L}{\rho v K_{\text{In}} C_P} * \frac{\mu C_P}{k}$$

which is the Group III Damkohler number Da^{III} times the Prandtl number Pr. Da^{III} describes the ratio of heat liberated or consumed to the bulk transport of heat and Pr describes momentum diffusivity to thermal diffusivity. Π_8/Π_7 simply tells us how K_{In} impacts ΔK_{IO}. And, Π_{10} provides information about the operating efficiency of the reactor.

The solution for scaling an adiabatic fixed-bed reactor is, by dimensional analysis

$$f\left(\frac{\Pi_1}{\Pi_2},\frac{\Pi_3}{\Pi_2},\Pi_2\Pi_4,\ \Pi_5,\frac{\Pi_1\Pi_6}{\Pi_5},\frac{\Pi_1^2\Pi_7}{\Pi_5},\frac{\Pi_1\Pi_6\Pi_9}{\Pi_4\Pi_7},\frac{\Pi_1^2\Pi_6\Pi_9}{\Pi_7},\frac{\Pi_8}{\Pi_7},\Pi_{10}\right)=0$$

Or, in terms of Π_{10} or C_{Out}/C_{In}, the solution is

$$\Pi_{10}=\kappa*f\left(\frac{\Pi_1}{\Pi_2},\frac{\Pi_3}{\Pi_2},\Pi_2\Pi_4,\ \Pi_5,\frac{\Pi_1\Pi_6}{\Pi_5},\frac{\Pi_1^2\Pi_7}{\Pi_5},\frac{\Pi_1\Pi_6\Pi_9}{\Pi_4\Pi_7},\frac{\Pi_1^2\Pi_6\Pi_9}{\Pi_7},\frac{\Pi_8}{\Pi_7}\right)$$

For the scaled fixed-bed reactor to operate similarly to the reference fixed-bed reactor, the following must hold

$$\left(C_{Out}/C_{In}\right)_{Scaled}=\left(C_{Out}/C_{In}\right)_{Reference}$$

and all the other dimensionless parameters must equal their counterparts. In other words,

$$\left(L/D\right)_{Scaled}=\left(L/D\right)_{Reference}$$

$$\left(d_P/D\right)_{Scaled}=\left(d_P/D\right)_{Reference}$$

$$Re_{Scaled}=Re_{Reference}$$

$$Sc_{Scaled}^{-1}=Sc_{Reference}^{-1}$$

$$\left(Lk_O/v\right)_{Scaled}=\left(Lk_O/v\right)_{Reference}$$

$$\left(L^2k_O/D_{Diff}\right)_{Scaled}=\left(L^2k_O/D_{Diff}\right)_{Reference}$$

$$\left((C_{In}\Delta H_R)Lk_O/\rho k v K_{In}\right)_{Scaled}=\left((C_{In}\Delta H_R)Lk_O/\rho k v K_{In}\right)_{Reference}$$

$$\left((C_{In}\Delta H_R)L^2k_O/kK_{In}\right)_{Scaled}=\left((C_{In}\Delta H_R)L^2k_O/kK_{In}\right)_{Reference}$$

$$\frac{\Delta K_{IO}}{K_{In}}\left(\Delta K_{IO}/K_{In}\right)_{Scaled}=\left(\Delta K_{IO}/K_{In}\right)_{Reference}$$

We know when starting to scale a fixed-bed reactor that

$$\left(d_P/D\right)_{Scaled}\neq\left(d_P/D\right)_{Reference}$$

Thus, geometric similarity between the scaled fixed-bed reactor and the reference fixed-bed reactor is not strictly held. The question is: can we accept this fact in our scaled design? If we are downscaling a fixed-bed reactor, the fluid flow velocity profile will be flat across the catalyst mass so long as $D > 10d_P$. If $D < 10d_P$, then a fluid flow velocity profile will develop across the catalyst mass. Thus

$$\left(C_{Out}/C_{In}\right)_{Scaled} \neq \left(C_{Out}/C_{In}\right)_{Reference}$$

due to a large fraction of the fluid slipping along the reactor wall and not contacting the catalyst mass *per se*. When upscaling from a small fixed-bed reactor, if $D < 10d_P$ for the reference, then

$$\left(C_{Out}/C_{In}\right)_{Scaled} \neq \left(C_{Out}/C_{In}\right)_{Reference}$$

due to a small fraction of the fluid slipping along the scaled reactor wall relative to the fluid slipping along the reference reactor wall. In this case, the result upon operating the scaled fixed-bed reactor will be

$$\left(C_{Out}/C_{In}\right)_{Scaled} > \left(C_{Out}/C_{In}\right)_{Reference}$$

In other words, the scaled fixed-bed reactor will perform better than the reference fixed-bed reactor.

It is unlikely that

$$(Re)_{Scaled} = (Re)_{Reference}$$

when we scale a fixed-bed reactor. The Reynolds number for laboratory-scale fixed-bed reactors and for pilot plant-scale fixed-bed reactors generally do not equal each other or the Reynolds number for a commercial-scale fixed-bed reactor. Therefore, different flow regimes exist at each reactor scale, which means the overall rate constant $k_{Overall}$ is different for each reactor scale. Thus

$$\left(Lk_O/v\right)_{Scaled} \neq \left(Lk_O/v\right)_{Reference}$$

$$\left(L^2 k_O/D_{Diff}\right)_{Scaled} \neq \left(L^2 k_O/D_{Diff}\right)_{Reference}$$

$$\left((C_{In}\Delta H_R)Lk_O/\rho k v K_{In}\right)_{Scaled} \neq \left((C_{In}\Delta H_R)Lk_O/\rho k v K_{In}\right)_{Reference}$$

$$\left((C_{In}\Delta H_R)L^2 k_O/k K_{In}\right)_{Scaled} \neq \left((C_{In}\Delta H_R)L^2 k_O/k K_{In}\right)_{Reference}$$

In this case, chemical similarity does not hold between the scaled fixed-bed reactor and the reference fixed-bed reactor. But, that does not mean we should abandon dimensional analysis. In fact, the opposite is true: we should use dimensional analysis to determine the extent of the chemical dissimilarity between two fixed-bed reactors. We do that by estimating $k_{Overall}$ for the scaled fixed-bed reactor and for the reference fixed-bed reactor from Figure 2.2, or from a similar relationship. We then insert the experimentally determined value into the above dimensionless ratios and compare each pair of dimensionless ratios. If heat required or released is our major concern, then we will want

$$
\begin{aligned}
\left((C_{In}\Delta H_R)Lk_O/\rho k v K_{In}\right)_{Scaled} &\approx \left((C_{In}\Delta H_R)Lk_O/\rho k v K_{In}\right)_{Reference} \\
\left((C_{In}\Delta H_R)L^2 k_O/k K_{In}\right)_{Scaled} &\approx \left((C_{In}\Delta H_R)L^2 k_O/k K_{In}\right)_{Reference}
\end{aligned}
$$

to be reasonably "close." If these two dimensionless parameters are not "close," then we will have to make a decision about continuing the project or we will have to make a design change. If the reaction under investigation is thermally neutral, then we will not be concerned with required or released heat; rather, we will be concerned with reaction performance. In other words, we want

$$
\begin{aligned}
\left(Lk_O/v\right)_{Scaled} &\approx \left(Lk_O/v\right)_{Reference} \\
\left(L^2 k_O/D_{Diff}\right)_{Scaled} &\approx \left(L^2 k_O/D_{Diff}\right)_{Reference}
\end{aligned}
$$

to be "close," if not equal. If these dimensionless parameters are not "close," then we will have to make a decision about continuing the project or we will have to make a design change.

It is difficult when scaling fixed-bed reactors to achieve geometric similarity because

$$
\left(d_P/D\right)_{Scaled} \neq \left(d_P/D\right)_{Reference}
$$

and it is difficult to maintain chemical similarity because

$$
(Re)_{Scaled} = (Re)_{Reference}
$$

which establishes different rate controlling regimes in different sized fixed-bed reactors. By using dimensional analysis, however, we can determine the extent to which a scaled fixed-bed reactor and its reference fixed-bed reactor are different. This information provides insight as to why

$$
\left(C_{Out}/C_{In}\right)_{Scaled} \neq \left(C_{Out}/C_{In}\right)_{Reference}
$$

occurs or forecasts that outcome. Knowledge of these differences saves much time, money, and finger-pointing during commissioning and the early operation of a scaled fixed-bed reactor. Time and money are saved because we are not trying to determine why

$$\left(C_{Out}/C_{In}\right)_{Scaled} \neq \left(C_{Out}/C_{In}\right)_{Reference}$$

We expected they would not be equal. And, since our expectation was met, the outcome becomes a nonevent.

5.6 SCALING PRESSURE DROP THROUGH A CATALYST MASS

When upscaling a solid-supported catalyst or when changing the size or shape of a catalyst to improve its performance, we must consider the impact that such a change will have upon the pressure drop of the fluid flowing through the fixed-bed reactor. Fixed-bed reactors are designed for a given pressure drop and the catalyst mass filling the reactor should not exceed that specification. If we are improving the performance of an existing process, then the new solid-supported catalyst cannot exceed the pressure drop generated by the current solid-supported catalyst. We know that increasing S_{Solid}/V_{Solid} increases the pressure drop generated by the catalyst mass; increasing A_P/V_P by reducing the size of the solid support also increases the pressure drop generated by the catalyst mass.

Traditionally, we have used the Ergun equation to estimate the pressure drop generated by a catalyst mass. Unfortunately, some engineers say the Ergun equation works well while many other engineers say it produces incorrect information. This disagreement generates doubt in the mind of those using the Ergun equation.

Ergun published his equation in 1952 and its purpose was to explain the transition zone between laminar flow through a mass of material particles and turbulent flow through the same mass of material particles. When we plot a function of pressure drop against the Reynolds number for the flowing fluid, we obtain a negatively sloped line for fluid flowing at low Reynolds numbers and a horizontal line for fluid flowing at high Reynolds numbers. We describe the former flow regime as laminar flow and the latter flow regime as turbulent flow. The experimental data curves smoothly from the laminar regime to the turbulent regime. The region between the laminar and turbulent flow regimes is the "transition" flow regime. Theoretically, the two flow regimes should make a sharp, obtuse intersection. Ergun derived an equation to explain the observed, smooth transition from the laminar flow regime to the turbulent flow regime.

We must derive the Ergun equation to understand its strengths and weaknesses. Consider a catalyst mass and imagine the interstitial space between its individual particles as forming small pipes or tubes through which fluid flows. If the fluid flowing through such a pipe or tube is at low Reynolds number, then from the Hagan−Poiseuille equation, the pressure drop Δp along the conduit is

$$\Delta p = \frac{8\mu v L}{r^2}$$

where μ is fluid viscosity; v is fluid velocity; L is pipe or tube length; and r is pipe or tube radius.

Pressure drop indicates flow efficiency. Inefficient flow has a large pressure drop while efficient flow has a small pressure drop. Flow efficiency depends on channel structure, which depends upon channel breadth and channel length. Unfortunately, we characterize flow by one channel variable, usually the radius or diameter of the flow area. We use "hydraulic radius" to better characterize a flow channel.[5] Hydraulic radius is defined as

$$r_{\mathrm{H}} = \frac{\text{Cross-sectional Flow Area}}{\text{Wetted Flow Perimeter}}$$

For circular pipes or tubes

$$r_{\mathrm{H}} = \frac{\pi r^2}{2\pi r} = \frac{r}{2}$$

Substituting hydraulic radius r_{H} for r in the Hagan−Poiseuille equation gives

$$\Delta p = \frac{8\mu v L}{4r_{\mathrm{H}}^2} = \frac{2\mu v L}{r_{\mathrm{II}}^2}$$

If spherical pellets comprise the catalyst mass, then their diameter provides a possible characteristic length to describe fluid flow around them. Multiplying r_H by L/L, where L is the length of the catalyst mass, gives

$$r_H = \frac{(\text{cross-sectional flow area}) * L}{(\text{wetted flow perimeter}) * L} = \frac{\text{flow volume}}{\text{wetted surface}}$$

$$= \frac{\varepsilon(\text{total volume})}{(1 - \varepsilon)(\text{wetted surface})}$$

where ε is the void fraction of the catalyst mass and $(1 - \varepsilon)$ is the solids fraction of the catalyst mass. Therefore, for spheres

$$r_H = \frac{(4/3)\pi r_S^3}{4\pi r_S^2} \frac{\varepsilon}{(1 - \varepsilon)} = \frac{r_S}{3} \frac{\varepsilon}{(1 - \varepsilon)}$$

where r_S is the radius of the catalyst spheres. Substituting into the Hagan–Poiseuille equation yields

$$\Delta p = \frac{2\mu v L}{(r_S/3)^2(\varepsilon/(1-\varepsilon))^2} = \frac{18\mu v L}{r_S^2} \frac{(1-\varepsilon)^2}{\varepsilon^2}$$

Converting to sphere diameter gives

$$\Delta p = \frac{72\mu v L}{d_S^2} \frac{(1-\varepsilon)^2}{\varepsilon^2}$$

where d_S is sphere diameter. Converting fluid velocity through the catalyst mass to fluid velocity through the empty reactor yields

$$\Delta p = \frac{72\mu v_\infty L}{d_S^2} \frac{(1-\varepsilon)^2}{\varepsilon^3}$$

where $v = (v_\infty/\varepsilon)$.

In the last equation above, we assumed a linear path through the catalyst mass, that path being given by L. But, no flow path through the catalyst mass is linear; i.e., straight; every path involves many twists and turns. We say the flow path is "tortuous." Ergun and others estimate that tortuosity to be

$$\left(\frac{25}{12}\right)L$$

Applying this adjustment to the above equation yields

$$\Delta p = \frac{72\mu v_{\infty}(25/12)L}{d_S^2}\frac{(1-\varepsilon)^2}{\varepsilon^3} = \frac{150\mu v_{\infty}L}{d_S^2}\frac{(1-\varepsilon)^2}{\varepsilon^3}$$

This last equation is the Blake–Kozeny equation, which is valid only for the laminar flow regime.[6]

The pressure drop for turbulent flow through a channel, such as a duct, is given as

$$\Delta p = \frac{f\rho L v^2}{2D}$$

where f is the friction factor for the flow; ρ is fluid density; L is channel length; v is fluid velocity; and D is channel breadth or diameter. As before, the channels through a catalyst mass possess irregular shapes; therefore, we must use hydraulic radius as the characteristic length to describe them. Assuming the channels formed through the catalyst mass have circular shapes allows us to use the above defined formula for hydraulic radius. Substituting into the above equation gives

$$\Delta p = \frac{f\rho L v^2}{8r_H}$$

But, for spherical catalyst pellets

$$r_H = \frac{d_S}{6}\frac{\varepsilon}{(1-\varepsilon)}$$

where d_S is the diameter of the spheres. Thus, the pressure drop becomes

$$\Delta p = \frac{f\rho L v^2}{(8/6)d_S(\varepsilon/(1-\varepsilon))} = \left(\frac{3f}{2}\right)\left(\frac{\rho L v^2}{2d_S}\right)\left(\frac{1-\varepsilon}{\varepsilon}\right)$$

Many studies concerning turbulent flow indicate that $3f/2$ is equivalent to 3.5. Making that substitution yields

$$\Delta p = 3.5\left(\frac{\rho L v^2}{2d_S}\right)\left(\frac{1-\varepsilon}{\varepsilon}\right)$$

Converting flow velocity through the catalyst mass to flow velocity through the empty reactor, then substituting into the above equation gives

$$\Delta p = 3.5\left(\frac{\rho L v_\infty^2}{2d_S}\right)\left(\frac{1-\varepsilon}{\varepsilon^3}\right) = 1.75\left(\frac{\rho L v_\infty^2}{d_S}\right)\left(\frac{1-\varepsilon}{\varepsilon^3}\right)$$

where v_∞ is the superficial fluid velocity through the reactor. This last equation is the Burke–Plummer equation for turbulent flow through a mass of spherical particles.[7]

The Blake–Kozeny equation and the Burke–Plummer equation define the limiting flow regimes for a fluid passing through a mass of spherical particles. Ergun added these equations to obtain

$$\frac{\Delta p}{L} = \frac{150\mu v_\infty}{d_S^2}\frac{(1-\varepsilon)^2}{\varepsilon^3} + 1.75\left(\frac{\rho v_\infty^2}{d_S}\right)\left(\frac{1-\varepsilon}{\varepsilon^3}\right)$$

Ergun then used available published data to confirm the validity of the equation. Note that the above equation is for spheres. For solid shapes other than spheres, we modify the equation by including a "sphericity" factor Ψ, which is defined as

$$\Psi = \frac{\text{surface area of sphere with equal volume to the nonspherical particle}}{\text{surface area of the nonspherical particle}}$$

The Ergun equation then becomes

$$\frac{\Delta p}{L} = \frac{150\mu v_\infty}{d_S^2 \Psi^2}\frac{(1-\varepsilon)^2}{\varepsilon^3} + 1.75\left(\frac{\rho v_\infty^2}{d_S \Psi}\right)\left(\frac{1-\varepsilon}{\varepsilon^3}\right)$$

Note that the Ergun equation

1. is not based on dimensional analysis, similitude, or model theory; therefore, it cannot be used to scale, either up or down;
2. depends upon the diameter of the catalyst spheres in the fixed-bed reactor but not on the diameter of the fixed-bed reactor, except through the calculation of v_∞;
3. assumes a tortuosity of 25/12;
4. assumes that three-halves of the friction factor equals 3.5;

5. contains two universal constants, namely, 150 and 1.75 that actually depend upon pellet or extrudate geometry and fixed-bed reactor geometry.[8]

In summary, it should not surprise us if the Ergun equation produces incorrect estimates of pressure drop for different shaped catalyst pellets and extrudates in the same fixed-bed reactor or that the Ergun equation cannot be used for scaling fixed-bed reactors.

We can perform a dimensional analysis of fluid flow through masses of material particles, such as catalyst masses, because the variables of the process are well known.[3] They have been studied for many decades. The variables are fluid velocity v; container diameter D, in our case, reactor diameter; characteristic length of the material mass L that is dependent upon the size and shape of the material particles; pressure drop per unit bed length Z; fluid density ρ; and, fluid viscosity μ. The dimension table for these variables is

Dimension/Variable	dP/Z	D	L	v	ρ	μ
L	-2	1	1	1	-3	-1
M	1	0	0	0	1	1
T	-2	0	0	-1	0	-1

We define L, the characteristic length of the material particle, as

$$L = \frac{Z}{\varepsilon}$$

where Z, in our case, is the height of the catalyst mass and ε is the void fraction of the catalyst pellets or extrudates. We determine Z from the amount of catalyst charged to the reactor, that amount being measured by weight $W_{Catalyst}$. Dividing $W_{Catalyst}$ by the loose compacted bulk density of the catalyst gives the volume of catalyst charged to the reactor, i.e.

$$\frac{W_{Catalyst}}{\rho_{LBD}} = V_{Catalyst}$$

We use loose bulk density of the catalyst rather than compacted bulk density of the catalyst because most commercial fixed-bed reactors are not shaken or vibrated during filling. Note that

$$V_{Cat} = A_{CS}Z$$

where A_{CS} is the cross-sectional area of the empty reactor. Dividing the above equation by the available flow area gives

$$\frac{A_{CS}Z}{\varepsilon A_{CS}} = \frac{Z}{\varepsilon} = L$$

Thus, if $Z = 3$ m and $\varepsilon = 0.4$, then $L = 4.5$ m; or, if $Z = 3$ m and $\varepsilon = 0.97$, then $L = 3.1$ m. In other words, L is an indication of pipe or tube length through the catalyst mass. Hence, L indicates the tortuosity of the flow path through the mass of material particles or through the catalyst mass.

We can write the above dimension table as a dimension matrix, which is

$$\begin{bmatrix} -2 & 1 & 1 & 1 & -3 & -1 \\ 1 & 0 & 0 & 0 & 1 & 1 \\ -2 & 0 & 0 & -1 & 0 & -1 \end{bmatrix}$$

The rank R for this dimension matrix is 3. Therefore, from Buckingham's Pi Theorem, the number of dimensionless parameters N_P for fluid flow through a mass of material particles will be

$$N_P = N_V - R$$

where N_V is the number of variables in the analysis and R is the rank of the dimension matrix. Making the appropriate substitutions gives

$$N_P = 6 - 3 = 3$$

Therefore, our dimensional analysis of fluid flow through a mass of material particles or through a catalyst mass will produce three dimensionless parameters. The total matrix for fluid flow through a catalyst mass is

$$T = \begin{array}{c} dP/dZ \\ D \\ L \\ v \\ \rho \\ \mu \end{array} \begin{bmatrix} 1 & 0 & 0 & 0 & 0 & 0 \\ 0 & 1 & 0 & 0 & 0 & 0 \\ 0 & 0 & 1 & 0 & 0 & 0 \\ -3 & 1 & 1 & -1 & -3 & -2 \\ -2 & 1 & 1 & -1 & -2 & -1 \\ 1 & -1 & -1 & 1 & 3 & 1 \end{bmatrix}$$

with column headers $\Pi_1\ \Pi_2\ \Pi_3$ over the first three columns.

The three independent dimensionless parameters are identified above the appropriate columns of the total matrix. The dimensionless parameters are

$$\Pi_1 = \frac{(dP/Z)\mu}{v^3 \rho^2}$$

$$\Pi_2 = \frac{Dv\rho}{\mu} = Re_D$$

$$\Pi_3 = \frac{Lv\rho}{\mu}$$

Π_1 is an unnamed dimensionless number; Π_2 is the Reynolds number based on the diameter of the catalyst mass or fixed-bed reactor; Π_3 is a dimensionless number resembling the Reynolds number, but based on the characteristic length of the catalyst mass. We can multiply and/or divide these three dimensionless variables with each other because they are independent of each other. Dividing Π_3 by Π_2 gives

$$\frac{\Pi_3}{\Pi_2} = \frac{(Lv\rho/\mu)}{(Dv\rho/\mu)} = \frac{L}{D}$$

which is an aspect ratio for the fixed-bed reactor. Therefore, the function describing fluid flow through a mass of material particles or through a catalyst mass is

$$f(\Pi_1, \Pi_2, \Pi_{3/2}) = 0$$

If we identify Π_1 as the dependent variable, then we can rewrite this function as

$$f(\Pi_2, \Pi_{3/2}) = \Pi_1$$

Thus, plotting Π_1 as a function of Π_2, the Reynolds number, produces smooth curves with parametric lines described by $\Pi_{3/2}$, which is shown in Figure 5.1. Figure 5.1 shows Π_1 as a function of Reynolds number for a variety of spherical particles in a number of different diameter pipes. $d(s)$ identifies sphere diameter. The curves are distinguished by the parametric $\Pi_{3/2}$, which is L/D. Figure 5.1 shows that Π_1 collapses to a common horizontal line in the turbulent flow regime, i.e., at high Reynolds numbers. This horizontal line corresponds to the Burke−Plummer result. For low Reynolds numbers, the correlation for each size sphere is negatively sloped, which corresponds to the

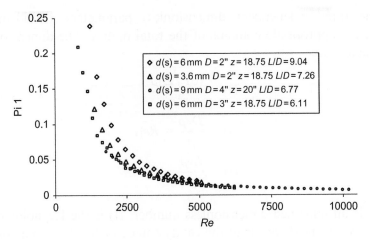

Figure 5.1 Dimensionless coefficient Π_1, or $(dP/Z)\mu/v^3\rho^2$, as a function of Π_2, or $(Dv\rho/\mu) = Re_D$, with $\Pi_{3/2}$, or L/D, as parameter.

Blake–Kozeny result. However, unlike the Ergun equation, the different sized spheres each produce a different correlation in the laminar flow regime. Each spherical diameter also produces a different correlation in the transition flow regime. Thus, the conflicting assessment with regard to the Ergun equation, namely, many engineers say it does not work while others say it does work. If fluid flows through a catalyst mass at high Reynolds numbers, it is more likely that the Ergun equation will produce valid results. If, however, fluid flows through a catalyst mass at low Reynolds numbers, then the Ergun equation is more likely to produce incorrect results. The same statement is valid for flow in the transition zone.

We can use the curves generated by dimensional analysis to scale fixed-bed reactors. First, plot $(dP/Z)\mu/v^3\rho^2$ against flow Reynolds number for a variety of catalyst pellets and extrudates. Then, decide whether to base the design on $(dP/Z)\mu/v^3\rho^2$ or on Reynolds number. If we base the design on Reynolds number, then read vertically upward from the chosen Reynolds number and record $(dP/Z)\mu/v^3\rho^2$ at all the intersections of the vertical line with the parametric L/D lines. Next, chose the $(dP/Z)\mu/v^3\rho^2$ appropriate for the design. Then calculate, using model theory, the diameter of the fixed-bed reactor from

$$\left(\frac{L}{D}\right)_{\text{Model}} = \left(\frac{L}{D}\right)_{\text{Prototype}}$$

where "prototype" designates the new design. Rearranging the above equation yields

$$D_{\text{Prototype}} = \left(\frac{L_{\text{Prototype}}}{L_{\text{Model}}} \right) D_{\text{Model}}$$

Scaling in this manner cannot be done using the Ergun equation.

5.7 SUMMARY

This chapter discussed fixed-bed reactor scaling and presented dimensional analysis as the preferred method for scaling such reactors. This chapter also discussed the use of dimensional analysis to scale fluid flow through a catalyst mass.

REFERENCES

1. Skoglund V. *Similitude: theory and applications.* Scranton, PA: International Textbook Company; 1967. p. 74–5.

2. Granger R. *Fluid mechanics.* New York, NY: Dover Publications, Inc.; 1995. p. 204.

3. Worstell J. *Dimensional analysis: practical guides in chemical engineering.* Oxford, UK: Butterworth-Heinemann; 2014 [chapter 7].

4. Tarhan M. *Catalytic reactor design.* New York, NY: McGraw-Hill Book Company; 1983. p. 80–1.

5. Janna W. *Introduction to fluid mechanics.* Monterey, CA: Brooks/Cole Engineering Division (Wadsworth, Inc.); 1983. p. 151.

6. Blake F. Resistance of Packing to Fluid Flow. *Trans Am Inst Chem Eng* 1922;**14**:415.

7. Burke S, Plummer W. Gas Flow Through Packed Beds. *Ind Eng Chem* 1928;**20**:1196.

8. Nemec D, Levec J. Flow Through Packed Reactors: I. Single-phase. *Chem Eng Sci* 2005;**60**:694.

9. Johnstone R, Thring M. *Pilot plants, models, and scale-up methods in chemical engineering.* New York, NY: McGraw-Hill Book Company, Inc.; 1957 [chapter 3].

CHAPTER 6

Regeneration of Solid-Supported Catalysts

6.1 INTRODUCTION

We use metal oxide solid supports, such as aluminas and silica−aluminas, as solid acid catalysts. Such solid-supported catalysts accumulate coke when placed in a fixed-bed reactor that processes hydrocarbon. The coke not only covers the initiating acid site, but it grows across the surface of the solid-supported catalyst, covering adjacent acid sites. Solid acids also undergo poisoning by Lewis bases. Catalysts comprising a metal deposited on a metal oxide solid support, such as those used for petroleum cracking, reforming, or hydrotreating, also experience poisoning, coke formation, sintering—both metallic cluster migration and pore surface area reduction—and pore plugging by metal sulfide deposition. We often call pore plugging "fouling." Thus, with time, the productivity of solid-supported catalyst declines, eventually reaching a point at which the product produced by the catalyst does not cover the cost of operating the process. When that occurs, the solid-supported catalyst in the fixed-bed reactor must be dumped and fresh solid-supported catalyst charged to the reactor or the solid-supported catalyst in the fixed-bed reactor must be regenerated. All these mechanisms deactivate solid-supported catalyst and we must know which mechanisms are operative when deciding whether to regenerate a solid-supported catalyst.

Solid-supported catalyst poisoning occurs reversibly or irreversibly. For reversibly poisoned solid-supported catalysts, streaming pristine feed through the catalyst mass desorbs poison from active sites, thereby restoring the activity of the catalyst. If pristine feed is not available, then, for metal oxides, streaming hot, dry nitrogen through the catalyst mass may volatilize poison off active sites. For zeolites, streaming mildly heated nitrogen through the catalyst mass volatizes poison off active sites. If reaction depends upon a reduced metal active site, we then must saturate the catalyst mass with hydrogen, to reduce the active site metal, prior to streaming it with feed.

Irreversible poisoning is not susceptible to regeneration. The conversion of a metallic active site to a metal sulfide is an example of irreversible poisoning. We may be able to regenerate such solid-supported catalysts in the laboratory, but the conditions will be so stringent as to make regeneration uneconomic at the commercial scale.

Sintering, both metal cluster migration and pore surface area reduction, are irreversible with regard to catalyst regeneration, although literature exists which suggests that metal clusters can be redispersed between 500°C and 600°C in oxidative environments.[1] Metal cluster redispersion requires experimental validation per solid-supported catalyst.

Pore plugging or fouling by metal sulfide deposition is also irreversible. The metal sulfide cannot be volatized from the pore nor can it be leached from the pore.

Our only recourse for reestablishing catalyst productivity when a solid-supported catalyst undergoes irreversible deactivation is to dump the effected, spent catalyst and charge fresh catalyst to the reactor.

Coke deposit, on the other hand, is fully amenable to regeneration via combustion. However, due to achievable combustion temperatures, sintering—both metal cluster migration and pore surface area reduction—does occur during regeneration of catalysts deactivated by coke accumulation.

6.2 TO REGENERATE OR NOT TO REGENERATE— THAT IS THE QUESTION

Economics dictates whether we regenerate deactivated solid-supported catalysts. Installing regeneration capability for a fixed-bed reactor is expensive. The reactor vessel, flanges, and associated piping must all be rated for the maximum achievable regeneration temperature, which is a combustion temperature in the case of regenerating coked catalyst. The exhaust gas may require catalytic oxidation to completely convert all the carbon to carbon dioxide. And, the emitted carbon dioxide may require permitting or may have to meet national "cap and trade" regulations. Also, the quantity of pipe and the number of valves required to regenerate a solid-supported catalyst *in situ* is significantly greater than that required for a dump and charge "regeneration" procedure. The *in situ* regeneration procedure is significantly more complex than the procedure for a dump and charge regeneration. And, the probability of and opportunity for a misstep increases as the complexity of an operating procedure grows. Such missteps can be dramatic in the case of *in situ* combustion regeneration.

If the solid-supported catalyst deactivates slowly, is inexpensive, is readily available, and contains nontoxic, nonhazardous metals, then dumping and charging fresh catalyst to the reactor will be the best economic option. Even if the catalyst deactivates relatively quickly, dumping and charging may provide the best economic option.

For solid-supported catalysts that contain precious metals, such as silver, platinum, palladium, or rhenium, the deactivated catalyst must be regenerated *in situ* or reclaimed via dumping and charging either fresh catalyst or off-site regenerated catalyst. Regenerating solid-supported catalyst off-site is expensive, but it may still be the better economic option when compared to the cost of *in situ* combustion regeneration.

6.3 REGENERATION OF COKED SOLID-SUPPORTED CATALYST

When we decide to regenerate the solid-supported catalyst contained in a fixed-bed reactor, we isolate the reactor from the process, then, if the process involves liquid hydrocarbon, we drain the reactor while flowing nitrogen top down through the catalyst mass. If the process involves hydrocarbon gas or vapor, we purge the catalyst mass top down with nitrogen. We slowly heat the catalyst mass with hot nitrogen to a temperature that will initiate coke combustion in the presence of oxygen after clearing hydrocarbon from the interstitial space of the fixed-bed reactor. At the specified temperature, we begin bleeding 0.5−1 mole percent oxygen into the hot nitrogen stream. Combustion begins when the oxygen encounters hot coke.

Coke combustion occurs in a narrow zone within the catalyst mass. Coke combustion is initially reaction rate limited since it occurs on the exterior surface of the catalyst pellet or extrudate. Combustion on the pellet or extrudate surface occurs at a constant temperature since heat is removed by convection, i.e., by the nitrogen flowing through the catalyst mass. As the coke burns, the combustion front enters the pores and begins to move toward the center of the catalyst pellet or extrudate. Temperature at the combustion front within the catalyst pellet or extrudate rises because heat is no longer removed by convection; rather, it diffuses along the pore. The increased temperature at the combustion front increases the reaction rate until finally combustion becomes oxygen diffusion rate limited. Combustion within the pores remains oxygen diffusion rate limited until all the coke is reacted or we stop the regeneration process. Note that coke combustion is independent of solid support structure, solid support history, and coke source.[1(p330)] Thus, neither pore size, PSD, nor the number of times a solid support has been regenerated impact coke regeneration.

We keep the nitrogen flow rate through the catalyst mass at a rate that separates the heat front from the combustion front. If the combustion and heat fronts coincide, the temperature of the event can, theoretically, become infinite. While not becoming infinite, the temperature at the combustion front will become high enough to induce sintering, both metal cluster migration and pore surface area reduction.[2]

As the heat front moves through the catalyst, it volatilizes any residual hydrocarbon in the pore of the solid support and it volatilizes the light components of the coke deposit. These volatilized hydrocarbons proceed downward through the catalyst mass. Eventually, these volatilized hydrocarbons enter the flare system and become flare gas.

Small amounts of transition metal oxide in the alumina or silica−alumina solid support increases the rate of regeneration. The presence of platinum in cracking and reforming catalysts thus accelerates their regeneration rate.[1(p332)] This technical area has become more heavily patented during the past 20 years, so conduct a "freedom to use" legal analysis before commercializing a regeneration procedure for a solid-supported catalyst.

The composition of coke varies, depending upon the source hydrocarbon and upon the time spent on the solid-supported catalyst at process temperature. Coke composition varies between $C_1H_{0.4}$ and C_1H_1. Thus, coke contains an appreciable amount of hydrogen and it oxidizes, i.e., burns, before carbon coke does, which means two reaction fronts exist during coke combustion: a hydrogen front and a carbon front. The hydrogen front precedes the carbon front.

A fair number of models have been proposed to describe coke combustion or catalyst regeneration.[2(pp218−242)] Such models become quite complex because they include pore diffusion rate limited hydrogen oxidation, pore diffusion rate limited carbon oxidation, and diffusion rate limited heat removal. The question becomes, does a complex model, which requires significant time and money to develop, provide us a better picture of reality than does a simple model, which we know from the start provides us a near-intuitive picture of reality? With regard to coke combustion, we are concerned with the removal of coke so as to restore catalyst productivity and with the temperature rise resulting from combustion so as to avoid sintering. In spite of the mechanistic complexity of coke combustion, oxidation kinetics are first-order in carbon and in oxygen. Thus, relatively simple models can be developed to provide the information we need concerning solid-supported catalyst regeneration.[3,4] However, each such model is catalyst and process dependent.

6.4 SUMMARY

This chapter presented the options available for regenerating solid-supported catalysts. The options were:

- dump and charge fresh catalyst;
- dump and charge catalyst regenerated off-site;
- regenerate catalyst *in situ*.

An economic analysis of each option must be done in order to make a selection. This chapter also discussed coke regeneration.

REFERENCES

1. Butt J, Petersen E. *Activation, deactivation, and poisoning of catalysts*. San Diego, CA: Academic Press, Inc.; 1988. p. 214–20.

2. Hughes R. *Deactivation of catalysts*. London, UK: Academic Press Inc.; 1984. p. 244.

3. Weisz P, Goodwin R. Combustion of Carbonaceous Deposits within Porous Catalyst Particles I. Diffusion-controlled Kinetics. *J Catal* 1963;2:397.

4. Weisz P, Goodwin R. Combustion of Carbonaceous Deposits within Porous Catalyst Particles II. Intrinsic Burning Rates. *J Catal* 1966;6:227.

Foundation of Dimensional Analysis

A.1 INTRODUCTION

Equations come in two varieties: mathematical and physical. Mathematical equations involve numbers that have no physical content, i.e., they involve pure numbers. We explore the relationships between pure numbers using the logic and rules of mathematics. We learn a fair number of these relationships during our mathematical preparation for an engineering career.

Scientists and engineers use physical equations. Physical equations are developed from experimental data and observation. They balance one set of physical magnitudes against another set of physical magnitudes via the equality sign of mathematics. The law for the conservation of energy is a good example of a physical equation. It was developed during the mid-nineteenth century through the effort of many scientists and engineers. For a flowing fluid, the physical concept for the mechanical conservation of energy is

Energy due to applied pressure		Kinetic energy		Potential energy		Energy loss due to friction		Shaft work per unit mass
	+		+		+		=	

The physical equation for the conservation of energy for a flowing fluid is, in the English Engineering system of units

$$\int \frac{dP}{\rho} + \Delta\left(\frac{\langle u \rangle^2}{2\alpha g_C}\right) + \left(\frac{g}{g_C}\right)\Delta z + F = -\frac{W_S}{m}$$

where P is pressure $[L^{-2}F]$; ρ is fluid density $[L^{-3}M]$; u is fluid velocity $[LT^{-1}]$—the bracket $\langle \rangle$ indicates an averaged value; α [dimensionless] describes the fluid flow profile within the conduit bounding the flowing fluid; g_C is the gravitational constant $[LMF^{-1}T^{-2}]$; g is gravitational acceleration $[LT^{-2}]$; z is height above the datum plane [L]; F is net frictional loss due to fluid flow $[LFM^{-1}]$; W_S is shaft work [LF]; and m is mass [M]. Thus, the dimension for each term is $[LFM^{-1}]$.

A.2 DEVELOPING DIMENSIONAL ANALYSIS

Since physical equations contain physical magnitudes, they must by necessity contain physical content. They contain physical content because physical magnitudes arise from physical quantities, which in turn arise from our perceptions. Therefore, when we write a physical equation, we are, in essence, writing an equation that balances physical quantities $\alpha[\Psi]$ through the use of an equality sign. Thus, we arrive at the first "axiom" of dimensional analysis.

Axiom 1: The numerical equality of a physical equation exists only when the physical magnitudes of that particular physical equation are similar, i.e., have the same units, which means the dimensions of the underlying physical quantities $\alpha[\Psi]$ are similar.[1]

In other words, a valid physical equation is dimensionally homogeneous, i.e., all its terms have the same dimensions and units.

All engineers and scientists learn this axiom upon their introduction to the study of Nature. We are told upon writing and solving our first physical equation that the individual terms of the given physical equation must have the same dimensions, i.e., units. We are also told that the dimensions and units of our calculated result must agree with the dimensions and units of the individual terms of the physical equation. For example, consider the physical equation

$$W = X - Y + Z$$

We can only calculate W if X, Y, and Z have the same dimensions and units. If X, Y, and Z each represent a physical magnitude of apples, then we can add and subtract them to obtain W, which will be a physical magnitude of apples. If X and Z have apple dimension and Y has orange dimension, then the above expression ceases to be a physical equation; it becomes meaningless from an engineering or scientific viewpoint.

Nonhomogeneous expressions do not contain physical information, thus they are not physical equations. The classic example of a non-homogeneous expression is

$$s + v = \frac{1}{2}at^2 + at$$

where s is distance [L]; v, velocity $[LT^{-1}]$; a, acceleration $[LT^{-2}]$; and t, time [T].[1(pp18–20),2] Writing this expression in dimensional terms gives us

$$[L] + [LT^{-1}] = [LT^{-2}][T^2] + [LT^{-2}][T]$$

which yields upon simplification

$$[L] + [LT^{-1}] = [L] + [LT^{-1}]$$

This last expression contains information, but that information does not describe a relationship between the left and right sides of the equality sign. No such relationship exists because the dimensions of the individual terms of the expression are mismatched. We frequently encounter nonhomogeneous expressions during our professional careers. Such expressions generally correlate, statistically, a product property to a process variable. In other words, the correlation describes a coincidence, not a cause and effect. Many such correlations exist in the polymer industry. Unfortunately, each such correlation is valid only for a given product from a particular production plant, which means the correlation possesses no physical information for another product or a different production plant.

We classify homogeneous physical equations as "restricted" and as "general." An example of a restricted equation is

$$s = (16.1)t^2$$

which describes the distance s [L] traversed by a free-falling object in time t [T]. Dimensionally, the above expression is

$$[L] = [T^2]$$

which makes it nonhomogeneous. However, we know, in certain situations, that it contains valid physical information. For this expression to be true, the coefficient 16.1 must have dimensions $[LT^{-2}]$. It, therefore, is not unreasonable for us to assume

$$16.1 = \frac{1}{2}g_0$$

where g_O is 32.2 ft/s^2 in the Old English Engineering system of units. Hence

$$s = (16.1)t^2$$

is a valid physical equation so long as the coefficient is a dimensional constant with Old English Engineering units. If this condition is true, the above expression becomes a restricted homogeneous physical equation. However, the above expression is not a physical equation if we use the SI system of units.

Now, consider Newton's Second Law

$$F = ma$$

It is an example of a general homogeneous physical equation since the dimensions on either side of the equality sign are $[LMT^{-2}]$. Its physical magnitudes can be expressed using any consistent system of units. Note that a general homogeneous physical equation does not contain a dimensional constant.[1(pp18–20)]

Consider our first ancestor who described to his fellow cave mates the concept of length and how to make a spear. To demonstrate how long to make a spear, he placed a straight, trimmed sapling on the cave floor and ensured that its larger end touched the cave wall. He then took his club and laid it beside the future spear, again ensuring that the end of the club touched the cave wall. Our ancestor then upended the club and walked it along the length of the future spear, counting each upending, until he reached its tip. Thus, our ancestor found the length of the future spear relative to the length of his club. Symbolically, he found

$$L_{Spear} = \alpha L_{Club}$$

where α is the number of times he upended the club from spear butt to tip. α is a pure number that we can manipulate with the logic and rules of mathematics. Note that L_{Spear} and L_{Club} are physical concepts, i.e., they are symbols and are not subject to the logic and rules of mathematics. Looking at his fellow cave conferees, our ancestor realizes that clubs come in a variety of lengths. So, he decides to step off the length of the future spear using his feet since most people have similar foot lengths. He, therefore, backed against the cave wall and began stepping heel-to-toe along the length of the future spear, then he did the same along the length of his club. He found that

$$L_{\text{Spear}} = \beta L_{\text{foot}}$$

and

$$L_{\text{Club}} = \gamma L_{\text{foot}}$$

Scratching his head, our ancestor realizes that the ratio of the future spear length to club length equals a pure number, namely

$$\frac{L_{\text{Spear}}}{L_{\text{Club}}} = \alpha$$

He realizes the same is true for his second measurement, hence

$$\frac{L_{\text{Spear}}}{L_{\text{Club}}} = \frac{\beta \, L_{\text{foot}}}{\gamma \, L_{\text{foot}}}$$

But the ratio of L_{Foot} is constant and can be deleted from this ratio. Thus

$$\frac{L_{\text{Spear}}}{L_{\text{Club}}} = \frac{\beta}{\gamma}$$

Equating the two ratios, our ancestor obtained

$$\frac{L_{\text{Spear}}}{L_{\text{Club}}} = \frac{\beta}{\gamma} = \alpha$$

Since α, β, and γ are pure numbers, our ancestor realized that the ratio of two physical quantities, in this case L_{Spear} and L_{Club}, is equal to the ratio of the numbers of units used to measure them, regardless of the system of units used to measure them.[3] In other words, the ratio of physical magnitudes of similar dimension is independent of the system of units. Thus, the ratio of physical magnitudes possesses an absolute significance independent of the system of units used to measure the corresponding physical quantity.[2(p19)]

Note that the above result makes it inherent that physical magnitude is inversely proportional to the size of the unit used, which is due to the linearity of our fundamental dimensions.[4] This result brings us to the second axiom of dimensional analysis, which states the following axiom.

Axiom 2: The ratio of physical magnitudes of two like physical quantities $\alpha[\Psi]$ is independent of the system of units used to quantify them, so long as the numerator and denominator of the ratio use the same system of units.[1(pp18–20)]

For example, let's assume our ancestor with the 50 ft by 100 ft garden plot has found a buyer for it. This buyer, unfortunately, lives in the neighboring kingdom where they measure length in "rods." The buyer has no idea what a foot length is and our ancestor has no idea what a rod length is. Therefore, the buyer brings his measuring rod to our ancestor's garden plot and finds it to be 3 rods by 6 rods.

The ratio of the length to breadth of our ancestor's garden plot is, in the English Engineering system of units

$$\frac{100 \text{ ft}}{50 \text{ ft}} = 2$$

and in rods the ratio is

$$\frac{6 \text{ rods}}{3 \text{ rods}} = 2$$

as per Axiom 2. Note that the resulting ratios are dimensionless. Dividing one ratio by the other yields

$$\frac{100 \text{ ft}/50 \text{ ft}}{6 \text{ rods}/3 \text{ rods}} = \frac{2}{2} = 1$$

Thus, we can equate the two ratios, namely,

$$\frac{100 \text{ ft}}{50 \text{ ft}} = \frac{6 \text{ rods}}{3 \text{ rods}}$$

which means that, within a given set of fundamental dimensions, all systems of units are equivalent. In other words, there is no distinguished or preferred system of units for a given set of fundamental dimensions.

We can also demonstrate Axiom 2 using a common engineering ratio. Consider the Reynolds number for fluid flow in a pipe, which is defined as

$$Re = \frac{\rho D v}{\mu}$$

where ρ is fluid density $[ML^{-3}]$; D is the pipe's diameter $[L]$; v is fluid velocity $[LT^{-1}]$; and μ is fluid dynamic viscosity $[L^{-1}MT^{-1}]$. In the English Engineering system of units, the density of water at $20°C$ is $62.3\ lb_M/ft^3$ and its viscosity is $2.36\ lb_M/ft*h$ or $0.000655\ lb_M/ft*s$. If the pipe's diameter is 1 foot and the water is flowing at 100 ft/s, then the Reynolds number is

$$Re = \frac{(62.3\ lb_M/ft^3)(1\ ft)(100\ ft/s)}{0.000655\ lb_M/\ ft*s} = \frac{6230\ lb_M/ft*s}{0.000655\ lb_M/ft*s} = 9.5 \times 10^6$$

In the SI system of units, water density at $20°C$ is $998\ kg/m^3$ and its dynamic viscosity is $0.000977\ kg/m*s$. The equivalent pipe diameter is $0.305\ m$ and the equivalent water flow rate is $30.5\ m/s$. The Reynolds number is then

$$Re = \frac{(998\ kg/m^3)(0.305\ m)(30.5\ m/s)}{0.000977\ kg/m*s} = \frac{9284\ kg/m*s}{0.000977\ kg/m*s} = 9.5 \times 10^6$$

We can equate the above two ratios

$$\frac{(998\ kg/m^3)(0.305\ m)(30.5\ m/s)}{0.000977\ kg/m*s} = \frac{(62.3\ lb_M/ft^3)(1\ ft)(100\ ft/s)}{0.000655\ lb_M/ft*s} = 9.5 \times 10^6$$

which shows that the English Engineering system of units is equivalent to the SI system of units. This result again suggests that no distinguished or preferred system of units exists for the length, mass, and time (LMT) set of dimensions.[5]

A.3 FOUNDATION OF METHOD OF INDICES

We can generalize this suggestion by considering a physical concept α that we want to quantify. Our first step is to choose a set of fundamental dimensions $[\Psi]$ that will quantify α. For example, let us choose LMT as our fundamental dimension set. We next select the system of units we will use to determine the physical magnitude of α. Since there are many such systems of units, let us choose $L_1M_1T_1$ as our system of units. Thus

$$\alpha[LMT] = \Phi(L_1, M_1, T_1)$$

where α represents a physical concept and $[\Psi]$ represents the fundamental dimensions quantifying α. $\Phi(L_1,M_1,T_1)$ represents the function determining the physical magnitude in the chosen system of units. We could have chosen a different system of units, which we identify as $L_2M_2T_2$. Note that $L_1M_1T_1$ and $L_2M_2T_2$ are related by a constant, β, which we have identified as a "conversion factor." Mathematically, the two systems of units are related as

$$\beta = \frac{\Phi(L_2, M_2, T_2)}{\Phi(L_1, M_1, T_1)}$$

Converting our physical quantity from the $L_1M_1T_1$ system of units to the $L_2M_2T_2$ system of units involves substituting $\Phi(L_2, M_2, T_2)/\beta$ for $\Phi(L_1, M_1, T_1)$; thus

$$\alpha[\text{LMT}] = \frac{\Phi(L_2, M_2, T_2)}{\beta}$$

Now, consider a third system of units designated $L_3M_3T_3$. Converting our physical quantity from the $L_1M_1T_1$ system of units to the $L_3M_3T_3$ system of units involves yet another conversion factor

$$\gamma = \frac{\Phi(L_3, M_3, T_3)}{\Phi(L_1, M_1, T_1)}$$

which upon substituting into

$$\alpha[\text{LMT}] = \Phi(L_1, M_1, T_1)$$

yields

$$\alpha[\text{LMT}] = \frac{\Phi(L_3, M_3, T_3)}{\gamma}$$

Dividing the last conversion by the previous conversion gives

$$\frac{\alpha[\text{LMT}]}{\alpha[\text{LMT}]} = \frac{\Phi(L_3, M_3, T_3)/\gamma}{\Phi(L_2, M_2, T_2)/\beta} = \frac{\beta\Phi(L_3, M_3, T_3)}{\gamma\Phi(L_2, M_2, T_2)} = 1$$

Thus

$$\frac{\gamma}{\beta} = \frac{\Phi(L_3, M_3, T_3)}{\Phi(L_2, M_2, T_2)}$$

Note that we could have done each of these conversions via a different route; namely, we could have converted each unit individually. Let us return to the conversion

$$\beta = \frac{\Phi(L_2, M_2, T_2)}{\Phi(L_1, M_1, T_1)}$$

and rearrange it. Doing so yields

$$\beta\Phi(L_1, M_1, T_1) = \Phi(L_2, M_2, T_2)$$

Dividing each term by its corresponding term in the first system of units, we get

$$\beta\Phi\left(L_1/L_1, M_1/M_1, T_1/T_1\right) = \Phi\left(L_2/L_1, M_2/M_1, T_2/T_1\right)$$

But

$$\Phi\left(L_1/L_1, M_1/M_1, T_1/T_1\right) = \Phi(1, 1, 1) = \kappa$$

where κ is a constant. Thus

$$\beta = \Phi\left(L_2/L_1, M_2/M_1, T_2/T_1\right)$$

Similarly for the third system of units

$$\gamma = \Phi\left(L_3/L_1, M_3/M_1, T_3/T_1\right)$$

Dividing the above two conversions gives

$$\frac{\gamma}{\beta} = \frac{\Phi\left(L_3/L_1, M_3/M_1, T_3/T_1\right)}{\Phi\left(L_2/L_1, M_2/M_1, T_2/T_1\right)}$$

Multiplying each term by its corresponding ratio of first system of units to second system of units gives

$$\frac{\gamma}{\beta} = \frac{\Phi\left(\frac{L_3 L_1}{L_1 L_2}, \frac{M_3 M_1}{M_1 M_2}, \frac{T_3 T_1}{T_1 T_2}\right)}{\Phi\left(\frac{L_2 L_1}{L_1 L_2}, \frac{M_2 M_1}{M_1 M_2}, \frac{T_2 T_1}{T_1 T_2}\right)} = \frac{\Phi\left(\frac{L_3 L_1}{L_1 L_2}, \frac{M_3 M_1}{M_1 M_2}, \frac{T_3 T_1}{T_1 T_2}\right)}{1} = \Phi\left(\frac{L_3 L_1}{L_1 L_2}, \frac{M_3 M_1}{M_1 M_2}, \frac{T_3 T_1}{T_1 T_2}\right)$$

Simplifying the above equation yields

$$\frac{\gamma}{\beta} = \Phi\left(L_3/L_2, M_3/M_2, T_3/T_2\right)$$

Equating the two γ/β equations gives us

$$\frac{\Phi(L_3, M_3, T_3)}{\Phi(L_2, M_2, T_2)} = \Phi\left(L_3/L_2, M_3/M_2, T_3/T_2\right)$$

Differentiating the above equation with respect to L_3 gives

$$\frac{\partial \Phi(L_3, M_3, T_3)/\partial L_3}{\Phi(L_2, M_2, T_2)} = \frac{1}{L_2} \Phi(L_3/L_2, M_3/M_2, T_3/T_2)$$

When we let $L_1 = L_2 = L_3$, $M_1 = M_2 = M_3$, and $T_1 = T_2 = T_3$, the above equation becomes

$$\frac{d\Phi(L, M, T)/dL}{\Phi(L, M, T)} = \frac{1}{L} \Phi(1, 1, 1) = \frac{a}{L}$$

where $\Phi(111)$ is a constant designated as "a." Rearranging the above equation gives

$$\frac{d\Phi(L, M, T)}{\Phi(L, M, T)} = a\left(\frac{dL}{L}\right)$$

Integrating yields

$$\ln(\Phi(L, M, T)) = a \ln(L) + \ln(\Phi'(M, T))$$

or, in exponential notation

$$\Phi(L, M, T) = L^a \Phi'(M, T)$$

where $\Phi'(M,T)$ is a new function dependent upon M and T only. Performing the same operations on M and T eventually produces

$$\Phi(L, M, T) = \kappa L^a M^b T^c$$

But, κ is a constant equal to 1; therefore

$$\Phi(L, M, T) = L^a M^b T^c$$

Thus, the dimension function which determines the physical magnitude is a monomial power law, as purported by Lord Rayleigh in 1877.[2(chp 2),5,6]

A.4 DIMENSIONAL HOMOGENEITY

We will now use the above result to prove Fourier's comments about dimensional homogeneity. Consider a dependent variable y represented by a function of independent variables x_1, x_2, x_3, ..., x_n. This statement in mathematical notation is

$$y = f(x_1, x_2, x_3, \ldots, x_n)$$

Let us assume the function is the sum of its independent variables, thus

$$y = x_1 + x_2 + x_3 + \cdots + x_n$$

If the function represents a physical equation, then each term in the function has a dimension associated with it; namely,

$$y[LMT] = x_1[L_1M_1T_1] + x_2[L_2M_2T_2] + x_3[L_3M_3T_3] + \cdots + x_n[L_nM_nT_n]$$

Substituting for y yields

$$(x_1 + x_2 + x_3 + \cdots + x_n)[LMT] = x_1[L_1M_1T_1] + x_2[L_2M_2T_2] \\ + x_3[L_3M_3T_3] + \cdots + x_n[L_nM_nT_n]$$

Expanding the terms to the left of the equality sign gives

$$x_1[LMT] + \cdots + x_n[LMT] = x_1[L_1M_1T_1] + \cdots + x_n[L_nM_nT_n]$$

Equating each term yields

$$
\begin{aligned}
x_1[LMT] &= \quad x_1[L_1M_1T_1] \\
x_2[LMT] &= \quad x_2[L_2M_2T_2] \\
&\vdots \qquad\quad \vdots \\
x_n[LMT] &= \quad x_n[L_nM_nT_n]
\end{aligned}
$$

But, from above

$$\alpha[\Psi] = \alpha[LMT] = \Phi(L, M, T) = L^a M^b T^c$$

Thus, the above set of linear equations becomes

$$
\begin{aligned}
x_1[LMT] &= \quad L^a M^b T^c = \quad L^{a_1} M^{b_1} T^{c_1} = \quad x_1[L_1M_1T_1] \\
x_2[LMT] &= \quad L^a M^b T^c = \quad L^{a_2} M^{b_2} T^{c_2} = \quad x_2[L_2M_2T_2] \\
&\vdots \qquad\qquad \vdots \qquad\qquad \vdots \qquad\qquad \vdots \\
x_n[LMT] &= \quad L^a M^b T^c = \quad L^{a_n} M^{b_n} T^{c_n} = \quad x_n[L_nM_nT_n]
\end{aligned}
$$

Removing the leftmost and rightmost terms since they are superfluous yields

$$
\begin{aligned}
L^a M^b T^c &= \quad L^{a_1} M^{b_1} T^{c_1} \\
L^a M^b T^c &= \quad L^{a_2} M^{b_2} T^{c_2} \\
&\vdots \qquad\qquad \vdots \\
L^a M^b T^c &= \quad L^{a_n} M^{b_n} T^{c_n}
\end{aligned}
$$

We can write the above set of equations more compactly as

$$L^a M^b T^c = L^{a_1} M^{b_1} T^{c_1} = \cdots = L^{a_n} M^{b_n} T^{c_n}$$

Equating like dimensions gives

$$L^a = L^{a_1} = \cdots = L^{a_n}$$
$$M^b = M^{b_1} = \cdots = M^{b_n}$$
$$T = T^{c_1} = T^{c_n}$$

which shows that, when adding, or subtracting, the dimension L, M, and T on each term must be the same.[7] In other words, we can only add apples to apples or oranges to oranges ... we cannot add apples and oranges to get "orpels."

A.5 MATRIX FORMULATION OF DIMENSIONAL ANALYSIS

Consider a dependent variable y represented by a function of independent variables $x_1^{k_1}, x_2^{k_2}, \ldots, x_n^{k_n}$, where the ks are constants. Mathematically

$$y = f(x_1^{k_1}, x_2^{k_2}, \ldots, x_n^{k_n})$$

If we assume the function is the multiplicative product of the independent variables, then

$$y = x_1^{k_1} x_2^{k_2} \ldots x_n^{k_n}$$

If the function represents a physical equation, then

$$y[LMT] = x_1^{k_1}[L_1 M_1 T_1]^{k_1} x_2^{k_2}[L_2 M_2 T_2]^{k_2} \ldots x_n^{k_n}[L_n M_n T_n]^{k_n}$$

Substituting for y in the above equation gives

$$(x_1^{k_1} x_2^{k_2} \ldots x_n^{k_n})[LMT] = x_1^{k_1}[L_1 M_1 T_1]^{k_1} x_2^{k_2}[L_2 M_2 T_2]^{k_2} \ldots x_n^{k_n}[L_n M_n T_n]^{k_n}$$

Then dividing by $(x_1^{k_1} x_2^{k_2} \ldots x_n^{k_n})$ yields

$$[LMT] = \frac{x_1^{k_1}[L_1 M_1 T_1]^{k_1} x_2^{k_2}[L_2 M_2 T_2]^{k_2} \ldots x_n^{k_n}[L_n M_n T_n]^{k_n}}{(x_1^{k_1} x_2^{k_2} \ldots x_n^{k_n})}$$
$$= [L_1 M_1 T_1]^{k_1}[L_2 M_2 T_2]^{k_2} \ldots [L_n M_n T_n]^{k_n}$$

But, as previously stated

$$\alpha[\Psi] = \alpha[LMT] = \Phi(L, M, T) = L^a M^b T^c$$

and

$$\alpha[LMT]^{k_n} = \Phi(L, M, T)^{k_n} = (L^a M^b T^c)^{k_n}$$

Making this substitution into the third equation above yields

$$L^a M^b T^c = (L^{a_1} M^{b_1} T^{c_1})^{k_1} (L^{a_2} M^{b_2} T^{c_2})^{k_2} \cdots (L^{a_n} M^{b_n} T^{c_n})^{k_n}$$

Equating the exponential terms for L, M, and T, respectively, gives

$$a = a_1 k_1 + a_2 k_2 + \cdots + a_n k_n$$
$$b = b_1 k_1 + b_2 k_2 + \cdots + b_n k_n$$
$$c = c_1 k_1 + c_2 k_2 + \cdots + c_n k_n$$

Note that above, we have n terms but only three equations. Therefore, to solve this system of linear equations, we need to assume or assign values to $n - 3$ terms. For convenience, let $n = 5$ in the above system of linear equations, then we have five unknowns and three equations; thus, we need to assume values for two unknowns. Let us assume we know k_3, k_4, and k_5. We will assume values for k_1 and k_2. We represent k_1 and k_2 as

$$k_1 = k_1 + 0 + 0 + \cdots + 0$$
$$k_2 = 0 + k_2 + 0 + \cdots + 0$$

Adding the k_1 and k_2 equations to the original set of linear equations gives us

$$k_1 = k_1 + 0 + 0 + \cdots + 0$$
$$k_2 = 0 + k_2 + 0 + \cdots + 0$$
$$a = a_1 k_1 + a_2 k_2 + \cdots + a_n k_n$$
$$b = b_1 k_1 + b_2 k_2 + \cdots + b_n k_n$$
$$c = c_1 k_1 + c_2 k_2 + \cdots + c_n k_n$$

which in matrix notation becomes

$$\begin{bmatrix} 1 & 0 & 0 & 0 & 0 \\ 0 & 1 & 0 & 0 & 0 \\ a_1 & a_2 & a_3 & a_4 & a_5 \\ b_1 & b_2 & b_3 & b_4 & b_5 \\ c_1 & c_2 & c_3 & c_4 & c_5 \end{bmatrix} \begin{bmatrix} k_1 \\ k_2 \\ k_3 \\ k_4 \\ k_5 \end{bmatrix} = \begin{bmatrix} k_1 \\ k_2 \\ a \\ b \\ c \end{bmatrix}$$

From matrix algebra, we can partition the above matrices into

$$\begin{bmatrix} 1 & 0 \\ 0 & 1 \end{bmatrix}$$

which is the identity matrix or unit matrix, represented by I, and the zero matrix

$$\begin{bmatrix} 0 & 0 & 0 \\ 0 & 0 & 0 \end{bmatrix}$$

represented by 0.[8] The matrix

$$\begin{bmatrix} a_1 & a_2 & a_3 & a_4 & a_5 \\ b_1 & b_2 & b_3 & b_4 & b_5 \\ c_1 & c_2 & c_3 & c_4 & c_5 \end{bmatrix}$$

is the dimension matrix. It follows directly from the dimension table. The dimension table catalogs the dimensions of each variable of the original function. Thus, the dimension table has the below format

Variable		x_1	x_2	x_3	x_4	x_5
Dimension	L	a_1	a_2	a_3	a_4	a_5
	M	b_1	b_2	b_3	b_4	b_5
	T	c_1	c_2	c_3	c_4	c_5

The dimension matrix can be partitioned into two matrices, one being a square matrix, i.e., a matrix with the same number of rows as columns; the other being the bulk, or remaining matrix elements. We define the square matrix as the rank matrix and the remaining matrix as the bulk matrix. Partitioning the above dimension matrix gives

$$\begin{bmatrix} \begin{bmatrix} a_1 & a_2 \\ b_1 & b_2 \\ c_1 & c_2 \end{bmatrix} & \begin{bmatrix} a_3 & a_4 & a_5 \\ b_3 & b_4 & b_5 \\ c_3 & c_4 & c_5 \end{bmatrix} \end{bmatrix}$$

We use the rank matrix to calculate the "rank" of the dimension matrix. We need the rank of the dimension matrix in order to determine the number of independent solutions that exist for our system of linear equations. From linear algebra, the rank of a matrix is the number of linearly independent rows, or columns, of a matrix.[9] In other words, the rank of a matrix is the number of independent equations in a system of linear equations. Thus, the number of variables in a system of linear equations, i.e., the number of columns in the dimension matrix minus the rank of the dimension matrix equals the number of selectable unknowns.[8] Mathematically

$$N_{\text{Var}} - R = N_{\text{Selectable}}$$

where R is the rank of the dimension matrix.

To determine the rank of the dimension matrix, we must calculate the determinant of the rank matrix. If the determinant of the rank matrix is nonzero, then R is the number of rows or the number of columns in the rank matrix. The above rank matrix is a 3×3 matrix; therefore, the rank of its dimension matrix is 3. In this case, $N_{\text{Var}} = 5$ and $R = 3$; therefore $N_{\text{Var}} - R = N_{\text{Selectable}}$ is $5 - 3 = 2$. Therefore, to solve the above set of linear equations, we need to select two unknowns.

We can now rewrite the matrix equation

$$\begin{bmatrix} 1 & 0 & 0 & 0 & 0 \\ 0 & 1 & 0 & 0 & 0 \\ a_1 & a_2 & a_3 & a_4 & a_5 \\ b_1 & b_2 & b_3 & b_4 & b_5 \\ c_1 & c_2 & c_3 & c_4 & c_5 \end{bmatrix} \begin{bmatrix} k_1 \\ k_2 \\ k_3 \\ k_4 \\ k_5 \end{bmatrix} = \begin{bmatrix} k_1 \\ k_2 \\ a \\ b \\ c \end{bmatrix}$$

in terms of the partitioned matrices; the above matrix equation becomes

$$\begin{bmatrix} I & 0 \\ B & R \end{bmatrix} \begin{bmatrix} k_1 \\ k_2 \\ k_3 \\ k_4 \\ k_5 \end{bmatrix} = \begin{bmatrix} k_1 \\ k_2 \\ a \\ b \\ c \end{bmatrix}$$

Its solution is[8]

$$\begin{bmatrix} k_1 \\ k_2 \\ k_3 \\ k_4 \\ k_5 \end{bmatrix} = \begin{bmatrix} I & 0 \\ -R^{-1}B & R^{-1} \end{bmatrix} \begin{bmatrix} k_1 \\ k_2 \\ a \\ b \\ c \end{bmatrix}$$

We define the total matrix to be

$$T = \begin{bmatrix} I & 0 \\ -R^{-1}B & R^{-1} \end{bmatrix}$$

With regard to dimensional analysis, the number of columns in the dimension matrix equals the number of variables in the system of linear equations and the difference between the number of columns in the dimension matrix and the rank of the dimension matrix equals the number of selectable unknowns in the system of linear equations. The number of selectable unknowns equals the number of columns in the identity or unit matrix I. The product of reading down a column of the identity matrix is a dimensional or dimensionless parameter, depending upon our selection of a, b, and c. If we select $a = b = c = 0$, the parameters will be dimensionless. If a, b, and c are nonzero, then the parameters will have dimensions. For the former case,

$$N_{Var} - R = N_P$$

where N_P is the number of independent dimensional or dimensionless parameters obtainable from a given set of linear equations. This result is known as Buckingham's Theorem or the Pi Theorem.[10-12] For the latter case,

$$N_{Var} - R + 1 = N_P$$

which is van Driest's rule, a variation of Buckingham's Theorem.[13]

A.6 IDENTIFYING VARIABLES FOR DIMENSIONAL ANALYSIS

The question always arises: how do we identify the variables for a dimensional analysis study? The best way to identify the variables for use in a dimensional analysis is to write the conservation laws and constitutive equations underpinning the process being studied. Constitutive equations describe a specific response of a given variable to an external force. The most familiar constitutive equations are Newton's law of viscosity, Fourier's law of heat conduction, and Fick's law of diffusion.

The issue when identifying variables for a dimensional analysis study is not having too many, but missing pertinent ones. In the former situation, we still obtain the correct result; however, that result will contain extraneous variables, variables not actually required by dimensional analysis. In the latter situation, dimensional analysis produces an incorrect result. Therefore, to ensure the correct result, we will include any variable we deem remotely pertinent to the process being investigated. We can then identify, during our analysis of the process, which variables are irrelevant.

We determine which variables are irrelevant by calculating the matrix V, which is

$$V = (-R^{-1} * B)^T$$

where superscript T identifies the "trace" of the resulting matrix $-R^{-1*}B$. A column of zeroes in matrix V identifies an irrelevant variable, a variable that can be dropped from the dimensional analysis of the process under study. We do not prove this assertion; it is proven elsewhere.[8(chps 8,10,11)]

A.7 SUMMARY

As with all engineering and scientific endeavors, dimensional analysis involves procedure. Procedures are mechanisms that help us organize our thoughts. They are outlines of what we plan to do. As such, they minimize the likelihood that we will overlook or ignore an important point of our project. In other words, procedures reduce the time we expend on a given project and increase the accuracy of our result.

The procedure for using the matrix formulation of dimensional analysis is

1. state the problem—clearly;
2. research all available literature for published results;
3. develop the pertinent balances, i.e., momentum, heat, and mass, for the problem;
4. list the important variables of the problem;
5. develop a dimension table using the identified variables;
6. write the dimension matrix;
7. determine the rank of the dimension matrix;
8. identify the rank matrix and calculate its inverse matrix;
9. identify the bulk matrix;
10. multiply the negative of the inverse rank matrix with the bulk matrix;

11. determine whether irrelevant variables are in the dimension matrix;
12. build the total matrix;
13. read the dimensionless parameters from the total matrix;
14. rearrange the dimensionless parameters to maximize physical content interpretation.

REFERENCES

1. Murphy G. *Similitude in engineering*. New York, NY: The Ronald Press Company; 1950. p. 17.

2. Bridgman P. *Dimensional analysis*. New Haven, CT: Yale University Press; 1922 [chapter 4].

3. Hunsaker J, Rightmire B. *Engineering applications of fluid mechanics*. New York, NY: McGraw-Hill Book Company, Inc.; 1947 [chapter 7].

4. Huntley H. *Dimensional analysis*. New York, NY: Dover Publications, Inc; 1967, page (First published by MacDonald and Company, Ltd., 1952).

5. Barenblatt G. *Scaling*. Cambridge, UK: Cambridge University Press; 2003. p. 17–20.

6. Barenblatt G. *Scaling, self-similarity, and intermediate asymptotics*. Cambridge, UK: Cambridge University Press; 1996. p. 34–7.

7. Langhaar H. *Dimensional analysis and theory of models*. New York, NY: John Wiley & Sons, Inc; 1951 [chapter 4].

8. Szirtes T. *Applied dimensional analysis and modeling*. 2nd ed. Burlington, MA: Butterworth-Heinemann; 2007 [chapter 7].

9. Jain M. *Vector spaces and matrices in physics*. New Delhi, India: Narosa Publishing House; 2001. p. 75.

10. Buckingham E. "On Physically Similar Systems; Illustrations of the Use of Dimensional Analysis". *Phys Rev* 1914;**4**(4):345.

11. Rayleigh L. "The Principle of Similitude". *Nature* 1915;**95**:66.

12. Tolman R. "The Specific Heat of Solids and the Principle of Similitude". *Phys Rev* 1914; 3:244.

13. Van Driest E. "On Dimensional Analysis and the Presentation of Data in Fluid-Flow Problems". *J Appl Mech* 1946;**13**(1):A-34.

Printed and bound by CPI Group (UK) Ltd, Croydon, CR0 4YY

08/05/2025

01864918-0001